CELLS

THE BUILDING BLOCKS OF LIFE

Cell Structure, Processes, and Reproduction

Cells: The Building Blocks of Life

CELLS
THE BUILDING BLOCKS OF LIFE

Cell Structure, Processes, and Reproduction

PHILL JONES

CHELSEA HOUSE
An Infobase Learning Company

CELL STRUCTURE, PROCESSES, AND REPRODUCTION

Copyright © 2011 by Infobase Learning

Chelsea House
An imprint of Infobase Learning
132 West 31st Street
New York NY 10001

Library of Congress Cataloging-in-Publication Data

Jones, Phill, 1953–
  Cell structure, processes, and reproduction / by Phill Jones.
      p. cm. — (Cells, the building blocks of life)
  Includes bibliographical references and index.
  ISBN 978-1-61753-004-3 (hardcover)
  1. Cell interaction—Popular works. 2. Cellular control mechanisms—Popular works.
  I. Title. II. Series.
  QH604.2.J66 2011
  571.6—dc22   2011004464

Chelsea House books are available at special discounts when purchased in bulk quantities for businesses, associations, institutions, or sales promotions. Please call our Special Sales Department in New York at (212) 967-8800 or (800) 322-8755.

You can find Chelsea House on the World Wide Web at http://www.infobaselearning.com

Text design and composition by Erika K. Arroyo
Cover design by Alicia Post
Cover printed by Yurchak Printing, Landisville, Pa.
Book printed and bound by Yurchak Printing, Landisville, Pa.
Date printed: August 2011
Printed in the United States of America

10 9 8 7 6 5 4 3 2 1

This book is printed on acid-free paper.

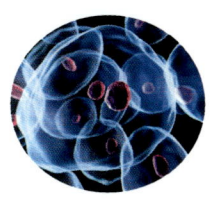

Contents

● ● ●

1

Introduction to Cells

Cells can live in very diverse environments. They live on land, in fresh water, and in the sea. They thrive in acid lakes, hot springs, volcanic vents at the bottom of the ocean, or in water ten times saltier than ocean water. Most cells live as single-celled organisms with sizes that vary from mycoplasma bacteria with a diameter of about 0.000007874 inch (0.2 micron) to seafloor-dwelling creatures with diameters of 1.18 inches (30 millimeters). Complex organisms, such as worms and whales, are multicellular life forms, which depend upon the coordinated activities of individual cells. At least 10^{13} cells collaborate in a wide variety of roles to maintain the structure and processes required by the human body.

Despite the diversity of cell types, cells generally share certain characteristics:

- A cell has the necessary information encoded in its genetic material to produce components for its structure, to perform activities necessary to sustain life, and to reproduce itself.
- One cell can reproduce itself by division, constructing two cells in its own image.
- A cell can gather energy and raw materials from the environment to perform a multiplicity of chemical reactions necessary to maintain life. A cell's series of life-sustaining chemical processes is the cell's **metabolism**.

- A cell can move materials within itself, and some cells can move within their environment.
- A cell can regulate its activities, such as metabolism.
- A cell can respond to a stimulus. For example, a cell may secrete a thick, protective coat around itself, modify its metabolism, or move to avoid potential danger.

CELLS DEPEND UPON FOUR TYPES OF LARGE MOLECULES

Cells can lose some of their general characteristics when they specialize to form the organs and tissues of multicellular organisms. Yet all cells share one feature: They require certain large molecules to sustain life. These molecules are **carbohydrates**, **lipids**, **nucleic acids**, and **proteins**.

Carbohydrates

Cells use carbohydrates as sources of energy and to construct various structural components. A carbohydrate is a compound consisting of carbon, hydrogen, and oxygen atoms. Usually, the three types of atoms occur in the ratio 1 carbon: 2 hydrogen: 1 oxygen. The atoms are linked with each other by **covalent bonds**. A covalent bond is a strong form of chemical bond in which two atoms share electrons.

Carbohydrates can be grouped into three classes: (1) **monosaccharides**; (2) **disaccharides**, consisting of two monosaccharides joined by a covalent bond; and (3) **polysaccharides**, which contain multiple sugar units. In general, cellular monosaccharides, or "simple sugars," have the molecular formula $(CH_2O)_n$, where n typically has a value of 3 to 7. For example, tetroses are sugars with four carbon atoms ($C_4H_8O_4$), pentoses have five carbon atoms ($C_5H_{10}O_5$), and hexoses have six carbon atoms ($C_6H_{12}O_6$). The hexose glucose, the most common monosaccharide, is an important energy-storing carbohydrate.

Polysaccharides are **polymers**. A polymer is a large chemical formed by combining smaller chemical units. Cellular polysaccharides are usually formed by combining glucose molecules into long chains. One example of a common polysaccharide is starch, the polymer used by most plants to store sugar. Animals store sugar as the polymer glycogen. Cellulose, another polysaccharide, is the main structural carbohydrate of plant cells and the major element of plant cell walls. It provides the woody structure of plants, and it is the most abundant organic compound on the planet.

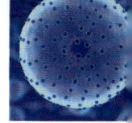

Lipids

The group of lipids includes a varied collection of **nonpolar biological molecules**. A nonpolar molecule is **hydrophobic**, which means "water fearing." Hydrophobic molecules do not dissolve in water and avoid interacting with water if possible. In contrast, **polar molecules** are **hydrophilic** ("water loving") molecules that both interact with and dissolve in water.

The three main types of cellular lipids are **neutral fats**, **phospholipids**, and **steroids**. Neutral fats provide an important source of fuel for animals. Cells use phospholipids to form vital structural components. Steroids are complex alcohols that have properties like fats. Examples of steroids include cholesterol and hormones, such as vitamin D.

Nucleic Acids

Cells contain two very important types of nucleic acids: **deoxyribonucleic acid (DNA)** and **ribonucleic acid (RNA)**. DNA contains coded instructions for synthesizing RNA and proteins. Several types of RNA molecules play vital roles in the production of proteins.

A DNA molecule is a polymer composed of **nucleotides** linked by covalent bonds. Each nucleotide has three parts:

- a **deoxyribose** sugar molecule, which is a five-carbon sugar molecule called **ribose** that is missing a particular oxygen atom;
- a chemical group that contains phosphorus; and
- a molecule called a **base**, which contains nitrogen.

The sugar group of one nucleotide binds with the phosphate group of another nucleotide to form a "sugar-phosphate-sugar-phosphate" structure, which is called the sugar-phosphate backbone of DNA. The bases of nucleotides stick out from the sugar-phosphate backbone. A DNA molecule has four types of bases: adenine, cytosine, guanine, and thymine. Scientists refer to the bases by the first letter of their names. For example, "AGCTGA" indicates a small piece of DNA that has the base sequence "adenine-guanine-cytosine-thymine-guanine-adenine."

An RNA molecule is similar to a DNA molecule, but RNA and DNA differ in three ways. First, RNA has a base called uracil that takes the place of thymine in DNA. For example, the sequence AGA TGT CCT in a piece of DNA would appear as AGA UGU CCU in an RNA molecule. Another difference between DNA and RNA is that DNA contains *deoxyribose*

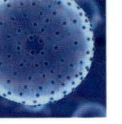
CHEMISTRY SETS

Thanks to the ingenious and ubiquitous chemistry set, children have learned about crystal formation, stink bomb fabrication, and other mysteries of chemical science. Home lab experimentation inspired generations of budding scientists.

The predecessor of the home chemistry set can be found in the eighteenth century's elaborate, portable chemical chests. Cabinets with chemical preparations and apparatus were assembled for physicians, druggists, mineralogists, and medical students, as well as for the amusement of the upper class. In the 1830s, manufacturers offered chemistry sets for children with a new element: the promise of magic. Young practitioners of chemical magic learned how to formulate disappearing inks and to transform a material's color.

From the 1860s to the 1940s, the British company John J. Griffin and Sons ruled the chemistry set market. The small shops and individuals who had produced traditional chemistry cabinets could not compete with the company's innovative tactic of versioning. The company sold 11 types of chemistry sets aimed at a spectrum of consumers from the casually curious to serious students receiving advanced instruction.

Two American companies introduced their chemistry sets for children during the early twentieth century. John J. Porter gets the credit for bringing the first chemistry sets to American children in 1915. The success of the company's simple kits, which cost 75 cents, inspired a competitor. Bolstered by the triumph of his Erector Set, Alfred Carlton Gilbert unveiled his Chemistry Outfit for Boys in 1917. The box bore the illustra-

sugars, whereas RNA contains *ribose sugars*. A third difference concerns the structure of RNA and DNA. RNA usually exists in the form of a single strand, whereas DNA can be found as a double-stranded helix.

Proteins

Proteins perform many structural functions in cells. Proteins also enable cells to execute activities necessary to sustain life. **Enzymes** are proteins that decrease the amount of energy required for a chemical reaction. This is vital for a cell. Otherwise, chemical reactions, such as metabolic activition of a solemn boy carefully mixing his chemicals, while the specter of a professional chemist hovered in the background. The company highlighted the gravity of the endeavor with the promise of a "career-building set," at a time when the American chemical industry enjoyed a rapid expansion.

During the 1940s through the 1950s, chemistry kit manufacturers declared that every house in America had one. Lionel, Skil-Craft, Chem-Pak, Midget Lab, and other competitors joined Porter and Gilbert. Porter's company offered 15 versions of its Chemcraft chemistry set and became the biggest consumer of test tubes. With their soaring popularity, companies sold increasingly elaborate chemistry sets. Premium Gilbert and Porter Chemcraft sets included balances, alcohol lamps, glassblowing apparatus, and dozens of chemicals. The Lab Technician Set for Girls, enclosed in a pink, zippered briefcase, joined the rugged, metal-encased Chemistry Outfit for Boys on store shelves.

The popularity of chemistry sets began to dissolve in the 1960s amidst new fears about science, environmental concerns, a focus on children's safety, and the deaths of chemistry set giants Gilbert and Porter. Government regulation of chemicals and litigation over alleged hazards of chemistry sets precipitated severe restrictions on the contents of a kit. Chemistry sets did enjoy a revival in the 1980s, but these sets differed from the classic kits that had been resplendent with scientific discovery and spiced with potential mischief. Arguably safer, the new chemistry sets offered peace of mind for parents at the expense of risk, adventure, and, some say, fun for young chemists.

ties, would require so much energy that high temperatures would disrupt cell structure.

In the process of **protein synthesis**, a protein polymer is formed by the addition of small molecules called **amino acids** that connect with each other by covalent bonds. Cells use 20 common types of amino acids and each amino acid has two basic components. One part, which is the same in all amino acids, allows amino acids to bind with each other. The other part—a side group—differs among the amino acids. One way to picture a protein is to imagine the identical parts of amino acids forming a chain.

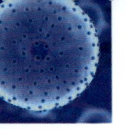

Each amino acid has a side group that sticks out from the chain. Some side groups are hydrophobic and move away from water, toward a protein's dry interior. Hydrophilic side groups move away from the protein's interior to the watery exterior of a protein. Other types of side groups are attracted to each other and bend the protein to move parts of the chain closer together. As amino acid side groups move to their preferred position, they bend and fold the protein.

A protein may have four types of structure that determine its final shape:

- The **primary structure** of a protein is the sequence of amino acids in the protein polymer.
- The **secondary structure** occurs when a protein's amino acids become linked by weak chemical bonds involving hydrogen atoms. Hydrogen bonds can pull one or more sections of a protein chain into the spiral shape of an alpha-helix or into a kinked structure, which is known as a beta-pleated sheet.
- The **tertiary structure** is formed as amino acid side groups bend, twist, and fold the protein into a complex three-dimensional shape. A protein's tertiary structure is stabilized by chemical bonds between pairs of amino acids located in different regions of the protein chain. For example, sulfur atoms in the side chains of two cysteine amino acids can form a bond called a disulfide bond.
- Some proteins have a **quaternary structure**. That is, the active protein molecule contains two or more protein subunits. A protein may have a number of identical subunits or a collection of different subunits.

The sequence of amino acids in a protein polymer defines the protein's primary structure. In turn, the primary structure determines secondary structure, tertiary structure, and the way that protein subunits interact to form a quaternary structure. The final shape of a protein is critical in determining the protein's function.

CELL ARCHITECTURE

Prokaryotes are one-celled organisms that do not develop into multicellular life forms. Bacteria, for example, are prokaryotes. A prokaryote cell can be divided into **cytoplasm** and a cell envelope. The cytoplasm is a gel-like matrix that fills the interior of the cell. A cell's metabolic processes,

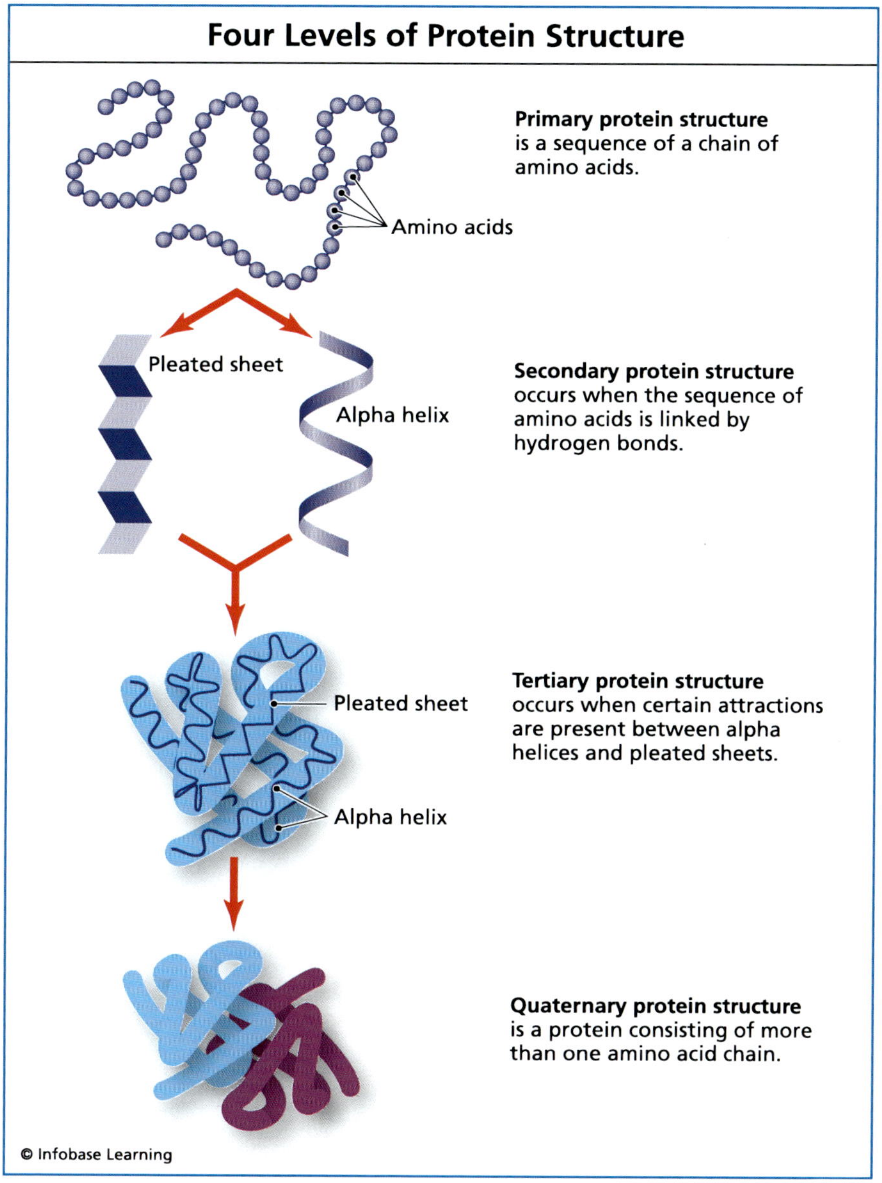

Four Levels of Protein Structure

Primary protein structure is a sequence of a chain of amino acids.

Amino acids

Pleated sheet

Alpha helix

Secondary protein structure occurs when the sequence of amino acids is linked by hydrogen bonds.

Pleated sheet

Alpha helix

Tertiary protein structure occurs when certain attractions are present between alpha helices and pleated sheets.

Quaternary protein structure is a protein consisting of more than one amino acid chain.

© Infobase Learning

FIGURE 1.1 There are four levels of protein structure. All proteins have a primary, secondary, and tertiary structure. Some have a quaternary structure as well.

which are required for life, take place in the cytoplasm. DNA, RNA, proteins, carbohydrates, and other important molecules can be found within the cytoplasm, as well as structures such as **ribosomes**, which play a vital

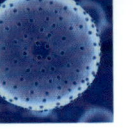

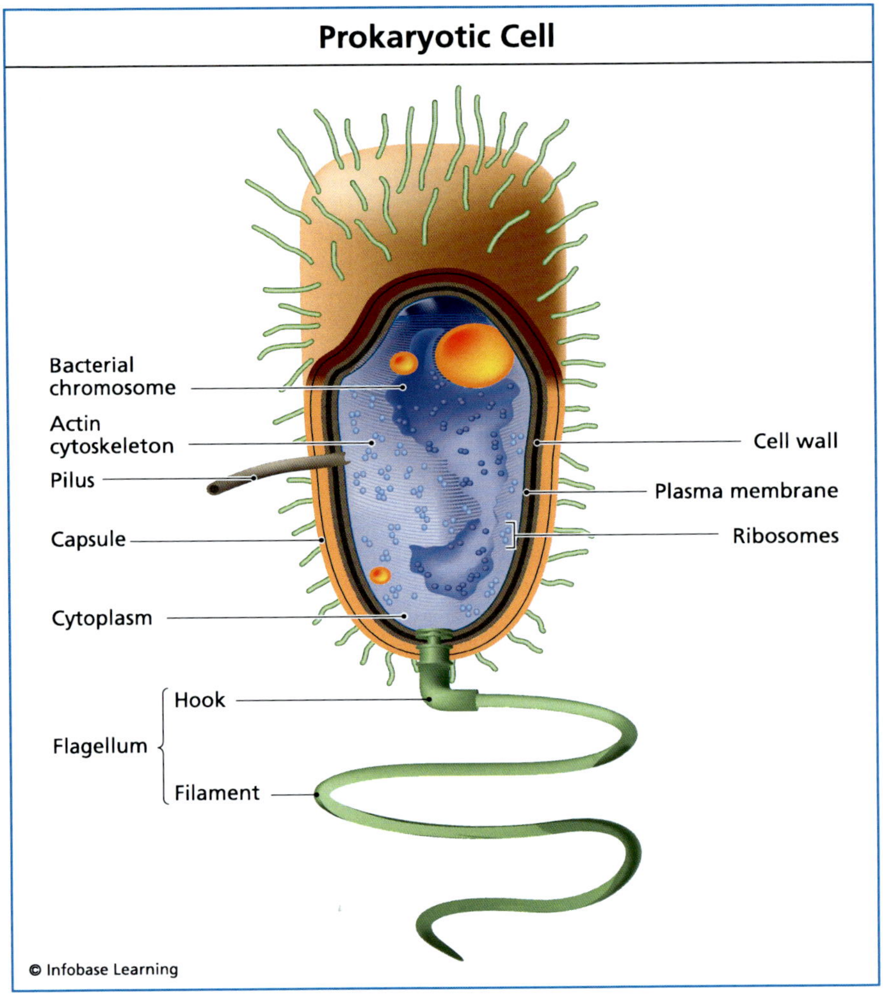

Prokaryotic Cell

Bacterial chromosome

Actin cytoskeleton

Pilus

Capsule

Cytoplasm

Cell wall

Plasma membrane

Ribosomes

Hook

Flagellum

Filament

© Infobase Learning

FIGURE 1.2 A single prokaryotic cell contains all of the necessary structures and components to carry out the processes necessary to sustain life and reproduce.

role in synthesizing proteins. Most prokaryotes have a single molecule of DNA, which contains the genetic information needed to produce proteins. This DNA molecule represents the cell's **genome**, the complete set of instructions for producing and maintaining the cell.

The cell envelope of most prokaryotes consists of at least two parts: a **plasma membrane** and a cell wall. The plasma membrane encloses the cytoplasm and controls the types of molecules entering and leaving a cell. A rigid cell wall encloses the plasma membrane and imparts a shape to the

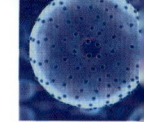

cell. Certain species also have a capsule, a coating of carbohydrates that covers the cell wall and can protect a prokaryote from dehydration.

Some prokaryote cells have appendages attached to the outside of the cell wall. Hair-like flagella move like propellers and enable cells to glide toward nutrients and away from danger. Some species have another type of hair-like structure called pili, which enable a cell to attach to a surface, such as a rock or a tooth.

Scientists propose that about two billion years ago, more complex cells evolved from prokaryotic cells. Over time, the plasma membrane extended into the cell to form compartments that divide tasks among small organ-like structures called **organelles**. One membrane compartment is the **nucleus**, a structure that contains the majority of a cell's DNA. A cell that has a nucleus is a **eukaryotic cell**.

A plasma membrane covers the outside of a eukaryotic cell and contains phospholipids, cholesterol, and **glycoproteins**, which are proteins with attached carbohydrates. The membrane retains a cell's contents, including a jellylike mix of water and proteins called **cytosol**. A cell membrane, however, is not simply a covering. Cell membranes regulate the flow of molecules into and out of the cell. A membrane allows some substances to pass freely into a cell, permits some to only pass with difficulty, and prevents other molecules from entering. As a selective barrier, a membrane enables a cell to accumulate nutrients from the environment, retain molecules for its own use, and excrete waste products.

Other structural features of a eukaryotic cell include the following:

- **Mitochondria** are jelly bean-shaped organelles that process molecules obtained from food to supply energy to the cell. Mitochondria are self-replicating and have a small DNA molecule that encodes some of the mitochondrial proteins.
- The **endoplasmic reticulum**, a collection of folded membranes, has two forms: rough endoplasmic reticulum and smooth endoplasmic reticulum. Rough endoplasmic reticulum—a site of protein synthesis—has ribosomes adhering to its outer surface, whereas smooth endoplasmic reticulum lacks ribosomes.
- **Golgi complexes** are collections of disk-shaped structures that aid in the delivery of the cell's proteins. Certain proteins are modified in Golgi complexes to prepare the proteins for export outside the cell.

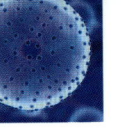

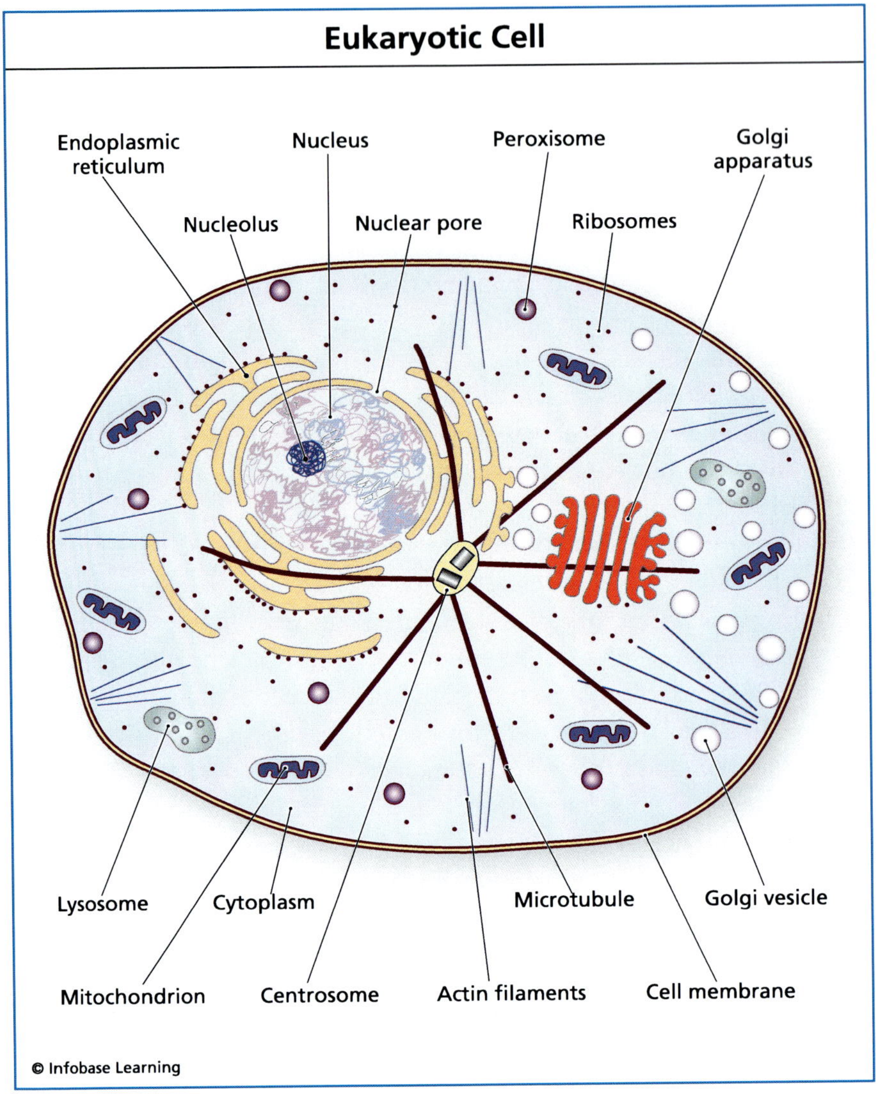

Eukaryotic Cell

Endoplasmic reticulum

Nucleus

Peroxisome

Golgi apparatus

Nucleolus

Nuclear pore

Ribosomes

Lysosome

Cytoplasm

Microtubule

Golgi vesicle

Mitochondrion

Centrosome

Actin filaments

Cell membrane

© Infobase Learning

FIGURE 1.3 In a eukaryotic cell, the nucleus contains the genome. The endoplasmic reticulum (ER) and the Golgi apparatus work together to modify proteins, many of which are destined for the cell membrane. These proteins are sent to the membrane in Golgi vesicles. Mitochondria provide the cell with energy. Ribosomes, some of which are attached to the ER, synthesize proteins. Lysosomes and peroxisomes recycle cellular material and molecules. The microtubules and centrosome form the spindle apparatus for moving chromosomes to the daughter cells during cell division.

GOING VIRAL

"H1N1 Is Still Spreading Globally," warned a report published in the October 26, 2009, edition of *The New York Times*. The article described a swine flu outbreak that President Barack Obama declared a national emergency. The H1N1 influenza virus caused the disease. Viruses plague all types of life forms, including bacteria, plants, and animals.

A virus consists of two basic components: a DNA or RNA genome wrapped in proteins. Viral genomes are too small to encode all proteins required for reproduction. A virus does not need to carry all of that data; a virus is a parasite. Some viruses reproduce by infecting an animal cell. The virus takes over the infected cell's functions to synthesize viral proteins and copies of the virus's genome, which assemble into viruses. After new viruses have formed, they can burst from an infected cell with such violence that they kill the cell. One infected cell can release as many as 100,000 virus particles.

Viruses that infect bacteria are called bacteriophages, or phages. The name "bacteriophage" means bacteria eater. Like viruses that infect animal cells, a phage is composed of a nucleic acid molecule—DNA or RNA—surrounded by proteins. Phages attach to bacteria and inject their genetic material. Once inside a bacterial cell, the phage's genetic material forces the bacteria's synthesis machinery to make more phages. In nature, phages play an important role in regulating the numbers of bacteria.

Viruses are not classified as cells. Most scientists do not even consider viruses to be alive because they lack a metabolic system, and they depend upon the cells that they infect to reproduce. In their 1983 dictionary, *Aristotle to Zoos*, scientists Peter and Jane Medawar wrote that, "No virus is *known* to do good: It has been well said that a virus is 'a piece of bad news wrapped up in protein.'"

- **Lysosomes** contain enzymes that digest old organelles, ingested food particles, and engulfed materials from the outside. These organelles break down complex chemicals to simple chemicals that the cell can use to make new products.

- **Peroxisomes** are organelles that purge toxic substances from a cell.
- The nucleus stores DNA that instructs the cell to make certain proteins and RNA molecules. Typically, the DNA can be found in **chromatin**, a mixture of DNA and proteins. Under the microscope, chromatin has a wiry, fuzzy appearance. When a cell is getting ready to reproduce itself, the chromatin compacts into the form of **chromosomes**. A membrane called the **nuclear envelope** surrounds the nucleus and separates the nucleus from other parts of the cell. Pores in the nuclear envelope allow RNA and other molecules to move outside the nucleus and into the cytoplasm. The inside of a eukaryotic cell can be considered to be divided into a nucleus and cytoplasm. Cytoplasm is simply the cytosol and organelles found outside the nucleus.
- Some eukaryotic cells, such as plant cells, have **chloroplasts** that use the energy of sunlight to synthesize carbohydrates from carbon dioxide and water in a process called **photosynthesis**. Like mitochondria, a chloroplast has a small DNA molecule and is self-replicating.

The cytoplasm contains another important component: the **cytoskeleton**, which provides structure (like a skeleton) and movement (like muscles). The cytoskeleton is composed of long proteins. These protein cables form tracks that allow molecules and organelles to move within a cell. The proteins also help to form extensions of the cell membrane that enable certain cells to travel.

CELL INFORMATION TRANSFER

Many consider the late twentieth century as marking the birth of the Information Age, a time when people can easily transfer and access information. Biologists could point to the evolution of the cell as a starting point for a different type of Information Age. Cells exist because they use a variety of mechanisms to transfer information from the environment, to the environment, and between cellular components. Information triggers a cell to reproduce itself, to alter its metabolism, or to avoid danger. Protein synthesis, one of the most important activities of a cell, depends on the transfer of information and the conversion of data from one form to another.

2

Information Transfer Within a Cell

All single-celled and multicellular organisms depend upon proteins for survival. For example, enzymes catalyze key chemical reactions of metabolism. Other proteins detect changes in a cell's environment and signal the cell to alter its activities, and still other proteins provide a structure for a cell. In a multicellular organism, feathers, horns, or hair are made of proteins. Proteins drive the contraction of animal muscles. Many types of animals use certain proteins to defend their bodies against invasion by bacteria and viruses.

The importance of proteins is shown by the ingenious system that cells use to produce protein polymers from amino acids. The system requires the transfer and transformation of data stored in DNA molecules.

A CLOSE LOOK AT DNA

Recall that DNA is a polymer of nucleotides and that each nucleotide has a base that sticks out from DNA's sugar-phosphate backbone. The four bases of a DNA molecule are adenine (A), cytosine (C), guanine (G), and thymine (T). In the nucleus of a cell, two strands of DNA form a double-stranded helix. This structure forms because certain bases are attracted to each other in a manner that can be imagined as a type of magnetic attraction. The rules of attraction are simple: An A on one DNA strand pairs with a T on the other DNA strand, and a G on one DNA strand pairs with a C on the other DNA strand. When bases of two different

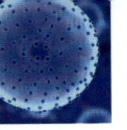

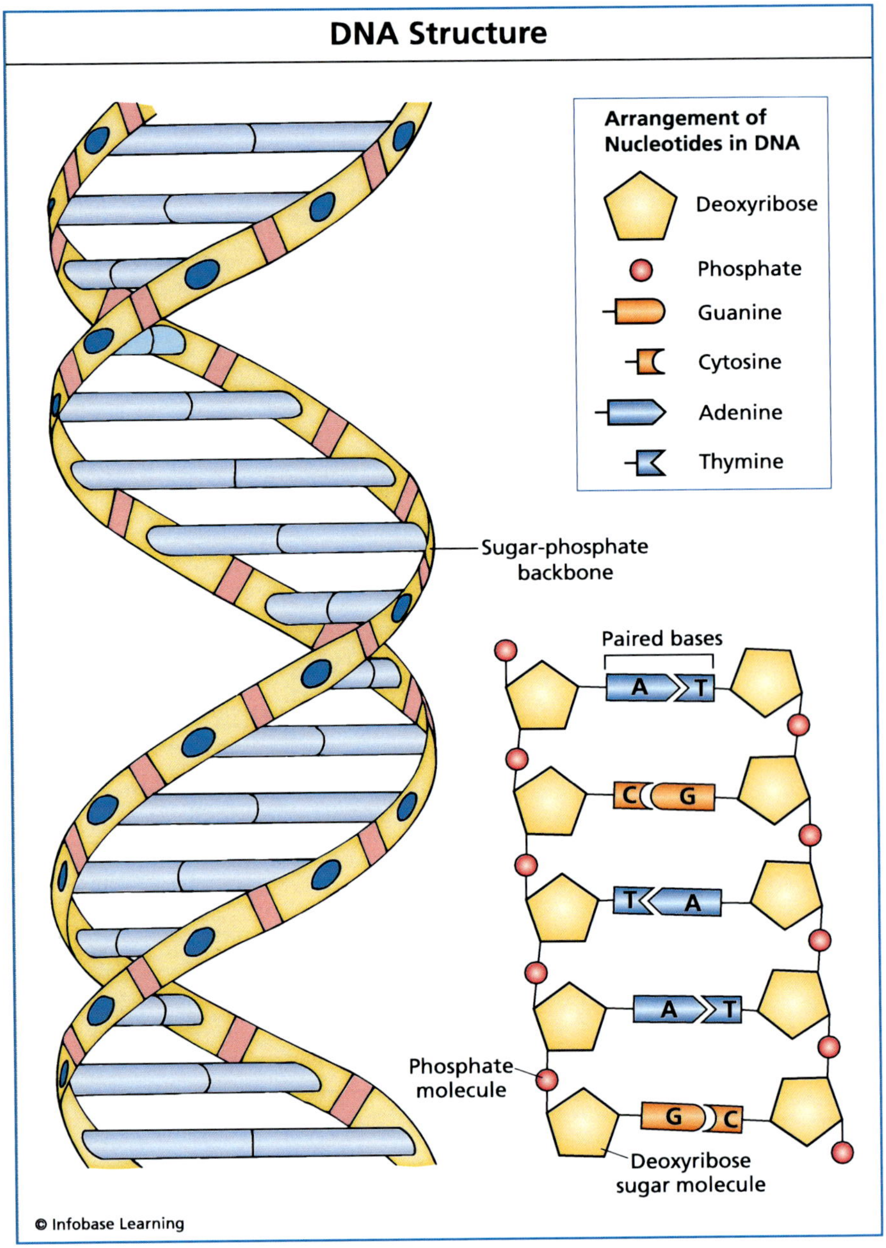

FIGURE 2.1 The structure of DNA resembles a ladder. The nucleotides twist in a double helix, joined together by the base pairs of nucleotides. The "rungs" of the ladder are made up of these base pairs.

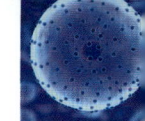

DNA strands bind together, they form a **base pair**. Consider a very short DNA molecule with two strands. One strand has the following sequence: CATTAGCATGGACT. The other strand would have the sequence GTA-ATCGTACCTGA. Together, the strands would appear as follows:

CATTAGCATGGACT

GTAATCGTACCTGA

This is the case because the first C in **C**ATTAGCATGGACT pairs with the first G in **G**TAATCGTACCTGA, the first A in C**A**TTAGCAT GGACT pairs with the first T in G**T**AATCGTACCTGA, and so on. The base pairs, AT and CG, have the same overall shape. Since they have the same shape, an AT base pair and a CG base pair can fit into any order between the two sugar-phosphate backbones without deforming the helix.

Instructions for producing proteins and certain RNA molecules can be found in the order of nucleotides in a DNA molecule. A **gene** is a DNA nucleotide sequence that provides the information a cell needs to synthesize a protein or RNA molecule. To produce a protein, the data stored in DNA must be transferred to the cell's protein production machinery.

TRANSFER OF GENETIC DATA FROM DNA TO RNA

The transfer of data from DNA is performed with molecules of RNA that are called **messenger RNA (mRNA)**. Messenger RNA has a nucleotide sequence that is a copy of a nucleotide sequence found in a DNA molecule. Of course, a messenger RNA molecule will not carry an exact copy of a DNA molecule's nucleotide sequence. DNA uses thymine, whereas RNA uses uracil.

Transcription is the synthesis of an RNA molecule from a DNA strand. In this process, an enzyme, called an **RNA polymerase**, travels along a DNA molecule, combining nucleotides to produce the RNA polymer. The enzyme incorporates nucleotides into the RNA molecule that are **complementary** to the nucleotides in a DNA strand, which is often called the template DNA strand. Nucleotides are complementary when they can form base pairs. For example, the following short DNA and RNA molecules have complementary nucleotide sequences:

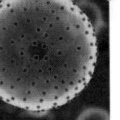

RNA: AGCUUAGCUAGGUUA

DNA: TCGAATCGATCCAAT

As the RNA polymerase advances through a section of the template DNA strand, the DNA double helix temporarily unwinds. The helix reforms after the enzyme has passed on to another section.

In prokaryotes, the RNA transcript serves as mRNA. Eukaryotic cells have a more complex and fascinating way to use the RNA transcript. The nuclei of eukaryotic cells contain much DNA that does not encode proteins. In humans, protein-encoded DNA amounts to about 2% of total DNA in a cell. Some of the noncoding DNA can be found in the middle of nucleotide sequences that encode a protein. As a result, a gene that encodes a protein has two types of DNA: **exons** (expressed regions), which are areas of DNA that encode parts of a protein; and **introns** (intervening regions), which are pieces of DNA that do not encode parts of a protein. A very simple gene may have a structure, such as:

[exon] – [intron] – [exon] – [intron] – [exon]

In a eukaryotic cell, transcription produces a pre-mRNA transcript that must be processed into mRNA. A vital part of the process is RNA splicing. During RNA splicing, introns are cleaved from pre-mRNA and exons are spliced together to produce an RNA molecule with a continuous stretch of protein-coding nucleotide sequences. As an example, the human dystrophin gene produces a **primary RNA transcript** that includes more than 80 introns, which are removed to produce the mature mRNA. The process of RNA splicing must be accurate. An error that adds or deletes just one nucleotide disrupts the nucleotide sequence that encodes a protein. A mistake made during this process can have huge consequences. For example, researchers discovered that one type of rickets, a disease caused by a lack of vitamin D, can be traced to an RNA splicing error that results in an inactive form of a protein that normally binds vitamin D.

Why do eukaryotic cells have such complex gene structures? According to one theory, the exon-intron arrangement allows cells to make different proteins by stitching together different combinations of exons during RNA splicing. The mixing of exons in diverse patterns is called **alternative splicing**. Alternative splicing explains a surprising discovery about human DNA. Human cells can make hundreds of thousands of different proteins. Yet, human DNA contains only about 20,000 protein-encoding

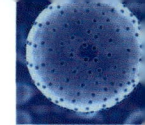

genes. Cells need fewer genes whenever RNA splicing can construct differ-ent mRNAs from the same primary RNA transcript.

THE GENETIC CODE

An mRNA molecule carries its recipe for a protein in the form of a **genetic code**. Mathematics provided an important clue about the nature of the genetic code. RNA contains only four different types of bases and proteins contain 20 common types of amino acids. This means that a single base does not code for a single amino acid. Suppose that the code used com-binations of two bases to signal one type of amino acid. In that case, the bases could code for only 16 different amino acids (4 x 4). If the genetic code used mixtures of three bases to code for an amino acid, then the bases would provide 64 combinations (4 x 4 x 4). This is more than enough to encode the 20 amino acids. A triplet code, therefore, would be the sim-plest type of code. During the 1960s, scientists verified that the genetic code is based upon base triplets.

A triplet of bases, which is called a **codon**, directs a cell to add an amino acid to produce a protein. The genetic code is based on the four nucleotide bases found in RNA: adenine, guanine, cytosine, and uracil. Some amino acids are encoded by two or more codons. For example, the amino acid leucine is encoded by the codons UUA, UUG, CUU, CUA, CUC, and CUG.

Not all codons stand for an amino acid; for example, some codons act like start-and-stop signals. Consider the following nucleotide sequence in a very small messenger RNA molecule:

. . . GCACGAUGGGGCGAAUUGCAUGCCCGUGA . . .

How should a cell's protein synthesis machinery read the sequence as triplets? There are three possibilities:

. . . GCA CGA UGG GGC GAA UUG CAU GCC CGU GA . . .

. . . GC ACG AUG GGG CGA AUU GCA UGC CCG UGA . . .

. . . G CAC GAU GGG GCG AAU UGC AUG CCC GUG A . . .

The codon AUG encodes the amino acid methionine. AUG also sig-nals the place in the nucleotide sequence of an mRNA where the code for a protein begins. The AUG codon creates the **reading frame** for the nucleotide sequence and determines how the sequence should be grouped into triplets. In the example, an AUG resides at nucleotides 6 through 8:

... GCACG**AUG**GGGCGAAUUGCAUGCCCGUGA ...

The AUG codon signals that the nucleotide sequence should be read as:

AUG GGG CGA AUU GCA UGC CCG UGA

Not every protein has a methionine amino acid at one end. The amino acid may be removed after protein synthesis has finished.

• • • BUSTED BY BACTERIA? • • •

Forensic specialists can find fingerprints and traces of DNA at a crime scene. Both types of evidence can identify a suspect. One day, forensic experts may find clues about the identity of a person who touched an object by analyzing bacteria left by that person's hands.

Noah Fierer and his colleagues at the University of Colorado in Boulder discovered that a typical human hand provides a home to a diversity of bacteria: about 150 species. The types of bacteria that live on skin vary greatly from person to person. Two people share only about 13% of bacterial species. In a 2008 study, the researchers identified more than 4,700 different bacteria species living on the hands of 51 people. Only five species dwelled on the skin of every participant of the study.

In 2010, Fierer announced that computer users leave DNA traces of bacteria on computer mice and keyboards. The DNA traces more closely match the DNA of bacterial colonies that inhabit the hands of the individual who used the computer, compared with bacterial DNA traces of randomly selected people. In one experiment, the researchers obtained useful samples of bacterial DNA two weeks after an individual touched a computer mouse.

The researchers are continuing their studies to determine if this type of bacterial DNA analysis will serve as a useful forensic tool. One application of the technique may be to identify which of two identical twins touched an object. Although identical twins share almost identical genomic DNA, the bacterial species that inhabit the skin of their hands differ.

The Genetic Code

Second letter

		U	C	A	G	
First letter	**U**	UUU UUC Phenyl-alanine UUA UUG Leucine	UCU UCC UCA UCG Serine	UAU UAC Tyrosine UAA Stop codon UAG Stop codon	UGU UGC Cysteine UGA Stop codon UGG Tryptophan	U C A G
	C	CUU CUC CUA CUG Leucine	CCU CCC CCA CCG Proline	CAU CAC Histidine CAA CAG Glutamine	CGU CGC CGA CGG Arginine	U C A G
	A	AUU AUC Isoleucine AUA AUG Methionine	ACU ACC ACA ACG Threonine	AAU AAC Asparagine AAA AAG Lysine	AGU AGC Serine AGA AGG Arginine	U C A G
	G	GUU GUC GUA GUG Valine	GCU GCC GCA GCG Alanine	GAU GAC Aspartic acid GAA GAG Glutamic acid	GGU GGC GGA GGG Glycine	U C A G

Third letter

© Infobase Learning

FIGURE 2.2 This table shows how different combinations of the four nucleotides in RNA encode different amino acids.

The genetic code signals the end of protein synthesis with a stop codon. In the example above, a UGA codon signals the end of a protein-encoding nucleotide sequence. UAA and UAG are also stop codons. Sometimes, the stop codons are referred to as "nonsense codons" because they code for nothing.

TRANSLATING THE GENETIC CODE INTO PROTEIN

In the next step of protein synthesis, genetic code data in an mRNA molecule is translated into a sequence of amino acids in a protein. Not surprisingly, scientists named this step **translation**. In eukaryotes, this step occurs after mature, spliced mRNA leaves the nucleus and enters the cytoplasm.

Translation depends upon molecules that ensure the assembly of the correct sequence of amino acids by binding an amino acid and its matching codon in mRNA. This critical function is performed by short, single-stranded RNAs called **transfer RNAs (tRNAs)**. One end of a tRNA molecule has an **anticodon**—three nucleotides that can form base pairs with a

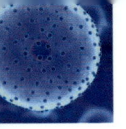

codon in mRNA. At its other end, the tRNA carries an amino acid speci-fied by the mRNA codon. A tRNA that carries its specified amino acid is often called a "charged tRNA." Each tRNA can bind one specific type of amino acid. Therefore, a cell must contain at least one tRNA for each of the 20 common amino acids. Enzymes known as tRNA transferases recognize the unique features of a tRNA and attach the correct amino acid. These enzymes are the key to the transfer of genetic data in protein synthesis. They read the language of the genetic code found at one end of a tRNA and match the code to a specific amino acid.

The genetic code has 61 nucleotide triplets that code for amino acids. Thanks to an effect known as **wobble**, a cell does not have to contain 61

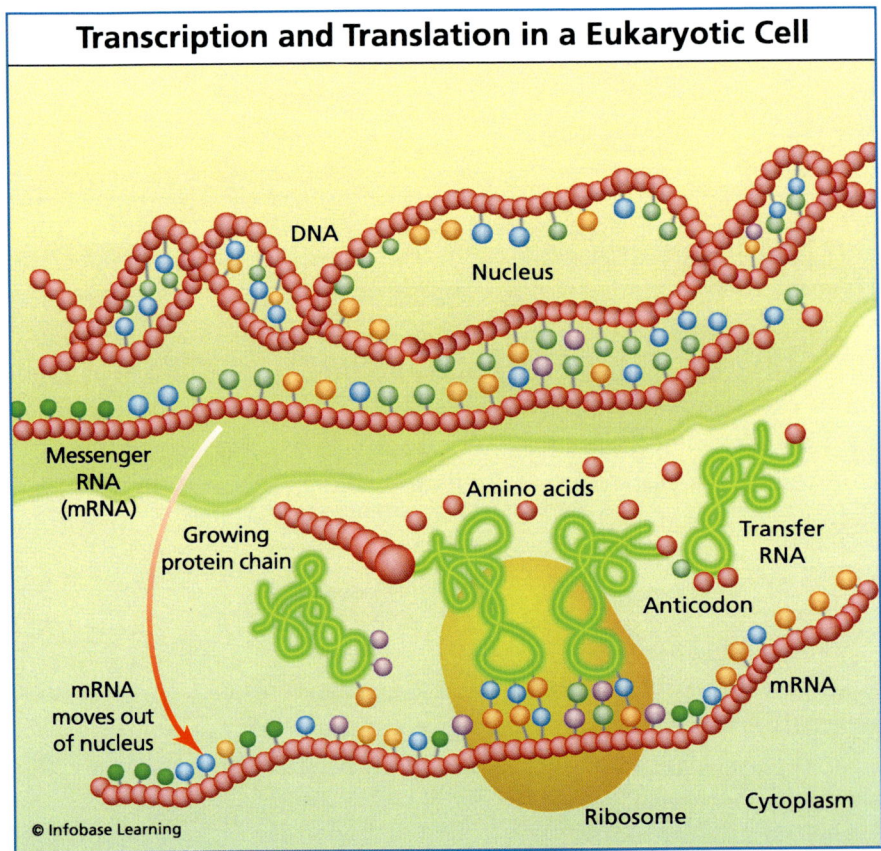

Transcription and Translation in a Eukaryotic Cell

DNA

Nucleus

Messenger RNA (mRNA)

Amino acids

Growing protein chain

Transfer RNA

Anticodon

mRNA moves out of nucleus

mRNA

© Infobase Learning

Ribosome

Cytoplasm

FIGURE 2.3 Transcription and translation are separate in eukaryotic cells. Transcription occurs in the nucleus to produce a pre-mRNA molecule. This molecule is typically processed to produce mature mRNA, which exists in the nucleus and is translated in the cytoplasm.

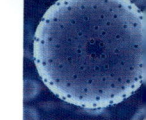

types of tRNAs. The wobble effect works like this: Certain amino acids are encoded by four codons that differ by only one nucleotide. The amino acid alanine, for example, is encoded by GCA, GCC, GCG, and GCU. For alanine, the third nucleotide of the codon adds nothing to the codon's specificity. That is, alanine is encoded by GCx, where x is any of the four nucleotides. There is a wobble at the third position. A single tRNA charged with alanine may recognize some or all of the four codons for alanine.

An mRNA molecule and charged tRNAs meet at ribosomes, which are among the most complex structures within a cell. Ribosomes are composed of RNA and many proteins. The ribosomes of bacteria contain more than 50 different types of proteins, whereas human ribosomes have about 80 different proteins. A ribosome recognizes the signal in mRNA for the start of translation. Ribosomes also stabilize interactions between mRNA and charged tRNAs and supply enzymatic activity that links amino acids from the tRNAs to form a protein. As ribosomes move along an mRNA molecule, they expose the mRNA's codons one by one to ensure correct addition of amino acids. After ribosomes reach a stop codon, they detach from the mRNA and the new protein.

PUTTING PROTEIN IN ITS PLACE

Plant and animal cells contain Golgi complexes, which are membrane-bound organelles. In animal cells, a typical Golgi complex contains five to eight membrane-covered pouches. If the nucleus is a cell's command center and the ribosomes of rough endoplasmic reticulum mark the production center, then the Golgi complexes are the shipping and distribution center for molecules produced by the cell.

After the completion of synthesis, newly made proteins become encapsulated in membrane-bound transport vesicles. The vesicles travel through the cytoplasm and fuse with membranes of a Golgi complex. Enzymes within the Golgi complex may remove segments of the protein or add molecules to the protein. As an example, many proteins are modified by the addition of carbohydrates to form glycoproteins. After processing, a protein is forced from the Golgi complex and, once again, travels in a membrane-bound vesicle. Depending upon certain chemical markers of a protein, the protein will be secreted from the cell, inserted into the plasma membrane, or positioned within a particular component of the cell.

As one example of a mechanism for protein distribution, consider proteins that are enzymes destined for the cell's lysosomes, which are

THE NEANDERTHAL IN YOU

DNA analyses of the human genome trace the human species to an ancestral population that migrated from East Africa throughout the rest of the world. As *Homo sapiens* explored the world, they met another type of humans: the Neanderthals. Anthropologists have found evidence that Neanderthals lived in Europe, the Middle East, and western Asia until about 30,000 years ago. Then, the Neanderthals seemed to disappear. However, new studies have revealed that the Neanderthals did not totally vanish; in fact, they live on in the DNA of humans today.

Svante Pääbo of the Max Planck Institute for Evolutionary Anthropology (located in Leipzig, Germany) led a research team who isolated DNA from three Neanderthal bones that are more than 38,000 years old. Using the DNA samples, they pieced together about 60% of a Neanderthal genome. Then, the researchers compared Neanderthal DNA sequences with DNA sequences of modern humans from different parts of the world. Their studies indicate that up to 4% of the genomic DNA of present-day humans who live outside Africa came from Neanderthals. In other words, early modern humans and Neanderthals interbred. The scientists suggest that, as *Homo sapiens* migrated through North Africa between 50,000 and 80,000 years ago, they interbred with Neanderthals. Rather than representing an experiment in a type of human that failed and died out, the Neanderthals are part of the inheritance of many living humans.

organelles that break down complex chemicals to simple chemicals. Lysosomal enzymes contain carbohydrate groups that are modified within the Golgi complex. After a Golgi complex enzyme recognizes a lysosomal enzyme, it adds a phosphate group to the carbohydrate. The combination of phosphate-sugar acts as a sorting signal that directs the transport of lysosomal enzymes from the Golgi complex within special vesicles. Having served their purpose, phosphate-sugar signals are cleaved from lysosomal enzymes and are returned to the Golgi complex for reuse. Meanwhile, the lysosomal enzymes are conveyed to lysosomes.

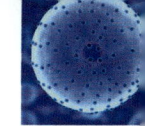

FROM DNA TO RNA TO PROTEIN

Nucleotide sequences in DNA molecules encode amino acid sequences of proteins. To synthesize a protein, a DNA sequence is transcribed into the nucleotide sequence of an mRNA molecule, which travels to the cell's ribosomes. Here, tRNA molecules carry amino acids to the mRNA, and, in effect, use the genetic code to translate the nucleotide sequence of an mRNA molecule into the amino acid sequence of a protein. A cell has systems to ensure that a protein is sent to its final destination: inside the cell, at the plasma membrane, or outside the cell.

A cell does not produce all proteins encoded in its DNA all the time. Cells use a variety of clever methods to control which proteins to produce and the amounts of a particular protein that must be produced.

3

Control of Gene Expression

A cell can make thousands of proteins. *Escherichia coli* (*E. coli*), for example, is a type of bacteria that has a genome that encodes about 4,300 proteins. A human cell carries about 20,000 protein-encoding genes. Yet cells do not produce all gene-encoded proteins at the same time. Such a massive effort would cause a type of gridlock as the cell quickly depletes chemicals and energy needed for protein synthesis. Cells of a multicellular organism must produce only certain proteins at a time for an additional reason: Different cells of the body have different functions. Selective protein synthesis enables nerve cells to transmit signals and muscle cells to power movement. The key to selective synthesis of proteins is control over **gene expression,** the process in which information stored in a DNA molecule is used to make a product.

TRANSCRIPTIONAL CONTROL: GENETIC SWITCHES

The Operon System of Bacteria

At any one time, bacteria such as *E. coli* produce a fraction of the proteins encoded by their genomes. The synthesis of many bacterial proteins depends upon the availability of the chemicals that are used for food sources. Consider the group of five genes that encode enzymes for production of the amino acid tryptophan (abbreviated trp).

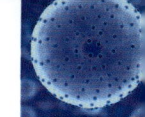

In the bacterial chromosome, the enzyme-encoded genes are clustered in a complex called an **operon**. A typical operon includes three basic parts:

- **structural genes** that encode proteins, such as enzymes
- a **promoter**, which is located in front of the structural genes; this is a segment of DNA where RNA polymerase binds to begin synthesis of mRNA (transcription)
- an **operator**, which may be located adjacent to the promoter or within the promoter, is a DNA segment where a protein called a repressor binds.

An operon can be pictured as follows:

[promoter] – [operator] – [structural genes]

A repressor protein can identify a particular nucleotide sequence in an operator and then bind tightly to the operator DNA segment. The binding of a repressor protein with its operator blocks RNA polymerase from producing mRNA of the structural genes.

RNA polymerase   →   →   →   →   →

[promoter] – [operator] – [structural genes]

RNA polymerase   →   repressor

↕

[promoter] – [operator] – [structural genes]

In the trp operon, a repressor protein alone cannot bind to the trp operator. If sufficient amounts of trp are available, trp binds with the repressor protein. The binding activates the trp repressor protein by changing its shape. An active trp repressor protein can bind with the trp operator. This is an efficient use of the cell's resources. If a bacterial cell cannot obtain trp from its environment, the cell must produce trp. In this state, the cell does not contain excess amounts of trp, which means that trp repressor proteins are free of trp and cannot bind with the trp operator. RNA polymerase is free to bind with the trp operon promoter and produce mRNA that encodes the enzymes for synthesizing trp. Suppose that a bacterial cell obtains sufficient amounts of trp from

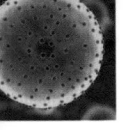

its environment. Now, excess trp molecules bind with trp repressor proteins and the active trp repressor proteins fasten to the trp operator and prevent transcription of the enzymes. The cell does not expend resources to make its own trp.

Since the trp operon can be repressed, it is called a *repressible operon*. Some operons can be stimulated, or induced, to function. This type of operon is known as an *inducible operon*. The lac operon is an example of an inducible operon, and is named after the sugar lactose. Lactose is a disaccharide that consists of glucose and galactose joined by a covalent bond. For lactose to be used as a source of energy, it must be broken down by a bacterial cell. The lac operon includes genes that encode an enzyme that is required to degrade lactose and a protein that increases the entry of lactose from the environment and into the bacterial cell. When lactose binds with the lac repressor protein, it deactivates the protein. That is, a lac repressor protein bound with lactose cannot bind with the lac operon. If a cell contains large amounts of lactose, the sugar binds with the lac repressor protein, which is fastened to the lac operon. The binding of lactose to the repressor protein alters the shape of the protein, causing it to release the lac operon. Once the lac operator is freed from the repressor protein, RNA polymerase can bind with the lac operon promoter and produce mRNA. The function of the lac repressor protein can be summarized as follows:

Excess lactose: lactose binds repressor $\rightarrow$ lactose-repressor complex releases operator

Low levels of lactose: repressor protein free of lactose $\rightarrow$ repressor binds operator

Why should the trp operon be repressible, whereas the lac operon is inducible? The trp operon encodes enzymes required to produce trp. The cell needs to turn off enzyme synthesis if plenty of trp is available. The lac operon encodes proteins needed to accumulate and break down lactose. The cell will need to synthesize proteins if lactose becomes available.

The repressor protein tactic is an example of a negative control of gene expression. Bacteria also use positive control strategies to regulate the expression of genes. As an example, certain proteins can bind near a promoter and assist the binding of RNA polymerase to increase the rate of transcription.

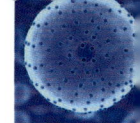

Genetic Switches in Eukaryotes

Like prokaryotes, eukaryotes regulate gene expression with proteins that affect the activity of RNA polymerase by recognizing certain nucleotide sequences in DNA and binding to those nucleotide sequences. However, eukaryotes have more complex mechanisms to control transcription. This is so, because eukaryotes have larger genomes, and in multicellular organisms, gene expression must be controlled to organize cells into tissues. Another important difference between prokaryotes, such as bacteria, and eukaryotic cells concerns the packaging of DNA.

Consider a human cell that has 46 chromosomes. The nucleus contains chromatin, a combination of DNA and proteins. In chromatin, twisted, double-stranded, negatively charged DNA spools around positively charged **histone** proteins to create a structure that is shaped like beads on a string. Each of these "beads," which are called nucleosomes,

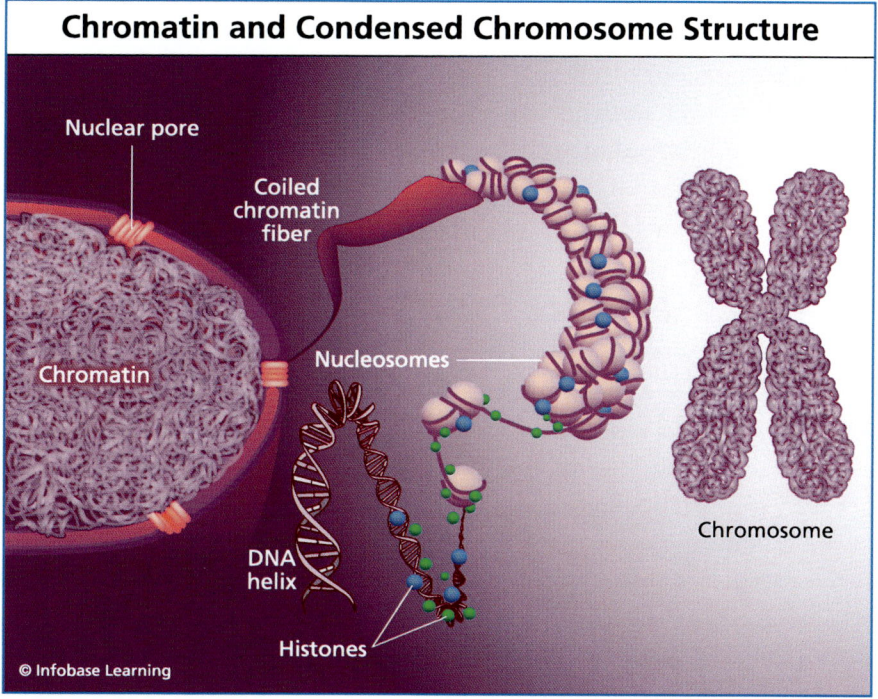

FIGURE 3.1 When DNA is combined with proteins, it organizes into a dense stringlike fiber called chromatin, which condenses into chromosomes during cell division. Each DNA strand wraps around groups of small protein molecules called histones, forming a series of beadlike structures called nucleosomes, which are connected by the DNA strand.

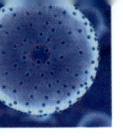

consists of 8 histones wrapped with a DNA segment of about 150 base pairs. DNA spacers of about 20 to 60 base pairs separate the nucleosomes. A human chromosome has hundreds of thousands to more than 1 million nucleosomes. The beads-on-a-string chromatin is compacted further into a dense, fiber-like structure called a solenoid. This compacting is necessary to squeeze 3 billion base pairs-worth of DNA into the cell nucleus.

The structure of chromatin affects the activities of genes. A dense, compact structure blocks RNA polymerase from gaining access to a gene's promoter for transcription. A relaxed, open structure of chromatin allows transcription to take place. The packaging of DNA into nucleosomes prevents transcription. As a result, the default of a eukaryotic gene is in the "off" position. Chromatin structure must be altered to allow RNA polymerase to access a promoter.

The DNA molecules of eukaryotic cells have many types of nucleotide sequences that regulate transcription. Some DNA control sequences reside near a promoter and assist in the binding of RNA polymerase. Other DNA control sequences may be located 50,000 base pairs or more from a promoter and can assist or hinder gene transcription. These distant DNA control sequences function because DNA loops in chromatin. A control sequence located 50,000 base pairs from a promoter when the DNA molecule is stretched into a straight line can be positioned close to a promoter when the DNA molecule is folded back onto itself.

TRANSCRIPTIONAL CONTROL: BEYOND GENETIC SWITCHES

Genetic switches operate because certain proteins, such as bacterial repressor proteins, bind with particular nucleotide sequences. Eukaryotic cells use several additional methods for controlling transcription that do not rely upon particular sequences of nucleotides. DNA methylation is an example of this type of control mechanism. A methyl group is a cluster of a carbon atom and three hydrogen atoms. In DNA methylation, an enzyme attaches a methyl group to a cytosine base in a DNA molecule. Attachment of methyl groups can hinder the synthesis of mRNA from DNA. By interfering with gene transcription, methylation stops the production of the protein encoded by the gene. DNA methylation is reversible: Certain enzymes can remove methyl groups from cytosine bases.

Another way to control gene expression involves proteins of chromatin. As previously discussed, chromatin contains double-stranded, negatively charged DNA that is spooled around positively charged histone

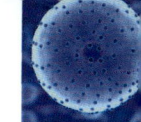

DNA and Histone Changes Affect Gene Activity

Gene "switched on"
- Active (open) chromatin
- Unmethylated cytosines (white circles)
- Histones with acetyl groups

———— Transcription possible ————

Gene "switched off"
- Silent (compact) chromatin
- Methylated cytosines (red circles)
- Histones without acetyl groups

———— Transcription impeded ————

© Infobase Learning

FIGURE 3.2 This image shows changes in chromatin organization that influence gene expression. Genes are expressed, or switched on, when chromatin is open, or active. They are inactivated, or switched off, when the chromatin is condensed.

proteins. Scientists have found at least 150 types of chemical alterations of histones. Some of these chemical changes affect whether chromatin is compact (inactive) or loose (active). One example of a chemical change is the addition of acetyl groups to histones. An acetyl group is a cluster of one oxygen (O) atom, two carbon (C) atoms, and three hydrogen (H) atoms. Enzymes attach an acetyl group to an amino group on a histone. An amino group has a positive electrical charge and contains a nitrogen (N) atom and three hydrogen atoms. The addition of an acetyl group ($CO\text{-}CH_3$) to an amino group (NH_3+) abolishes the amino group's positive charge ($NH\text{-}CO\text{-}CH_3$). Since the attachment of acetyl groups neutralizes the positive charge of histones, the attraction weakens between histones and negatively charged DNA. The weakened attraction causes the chromatin structure to unravel and allows transcription to take place. The reverse also happens: If an enzyme removes acetyl groups from histones, the proteins become more positively charged and bind more tightly to DNA. Removing acetyl groups can prevent transcription from a gene.

DNA methylation and chemical changes in histones often occur together. Inactive genes can be found in methylated regions of DNA with dense chromatin and histones that lack acetyl groups. Active areas of chromatin can have unmethylated DNA and large amounts of histones with acetyl groups. These mechanisms add a level of control to ensure that the correct genes are active in certain cells at the right times.

POST-TRANSCRIPTIONAL CONTROL OF GENE EXPRESSION

Recall that eukaryotic cells have genes that contain exons (areas of DNA that encode parts of a protein) and introns (areas of DNA that do not encode parts of a protein). To produce mRNA, nucleotide sequences of introns are removed and nucleotide sequences of exons are spliced together. In alternative splicing, various combinations of exons are stitched together to produce different mRNAs. Alternative splicing creates mRNAs for different proteins from the same gene. Alternative splicing also regulates the production of certain forms of a protein. Differently spliced mRNAs produce proteins that can have different locations within a cell and different functions. In multicellular organisms, the preferred form of a protein can depend upon the particular function of a cell or the stage of development of a tissue.

The expression of the human calcitonin gene shows the effect of alternative splicing. The primary RNA transcript of the gene, the "pre-mRNA," contains six exons (E) and five introns (I):

[E1] – [I1] – [E2] – [I2] – [E3] – [I3] – [E4] – [I4] – [E5] – [I5] – [E6]

In cells of the thyroid gland, 95% of the RNA is processed to include exons 1 through 4, producing a 32-amino-acid, calcitonin peptide. (A **peptide** is simply a short chain of linked amino acids that is too small to qualify as a protein.) In certain nerve cells, 99% of the RNA is processed to exclude exon 4 and include exons 5 and 6. This results in the 37-amino-acid, calcitonin gene-related peptide.

Calcitonin: [E1] – [E2] – [E3] – [E4]

Calcitonin gene-related peptide: [E1] – [E2] – [E3] – [E5] – [E6]

Calcitonin lowers the amount of calcium in the blood, whereas calcitonin gene-related peptide can stimulate nerve cells.

RNA interference, or RNAi, controls protein synthesis by destroying mRNA or by blocking translation. One mechanism of RNAi begins with

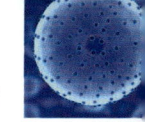

WHO LET THE DOGS OUT?

Humans have had a long partnership with dogs and have bred them for a variety of roles. Although the majority of today's 300 dog breeds were developed as recently as the Victorian era, ancient breeds of dogs first appeared more than 14,000 years ago. Who domesticated the first dogs? Scientists have debated about whether the first dogs were bred from their wolf ancestors in the Middle East or in China. In 2010, a massive DNA study answered the question: the Middle East, the birthplace of domesticated cats, many types of livestock, and agriculture.

In the study, researchers analyzed 48,000 DNA markers in the genomes of more than 900 dogs from 85 breeds. They also examined the DNA markers in the genomes of more than 200 wild gray wolves—the ancestor of domestic dogs—from 11 different populations around the world. The results showed that dogs share more similarity in their genomes with Middle Eastern gray wolves than with wolf populations that live in other parts of the world. This is consistent with the theory that dogs originated in the Middle East.

Other canine studies focus on the genetic basis for the diversity of dogs' characteristics. So far, the results point to a simple picture. For instance, variations in one gene can account for the short legs that are characteristic of certain dog breeds, such as dachshunds and corgis. Another gene can explain more than 50% of variation in body sizes among dog breeds. Researchers also have identified genes that produce fur types, and coat colors and patterns.

the synthesis of RNA molecules that have nucleotide sequences identical to parts of a gene. Enzymes cut the RNA molecules into pieces about 20 to 30 nucleotides long. The tiny RNA molecules are called **micro-RNAs**. Scientists propose that the cells of plants, fungi, and animals produce thousands or tens of thousands of different micro-RNAs. These micro-RNAs may play a role in controlling at least 70% of a cell's protein-encoding genes.

Micro-RNAs can inhibit the synthesis of a protein in two ways. In both cases, a micro-RNA binds to a multi-protein structure called the RNA-induced silencing complex, or RISC. When it finds its target, mRNA, micro-RNA-RISC attaches to the mRNA by base pairing with the micro-RNA. Burdened with its micro-RNA-RISC passenger, the mRNA gamely loads onto ribosomes for translation. However, the presence of one or more micro-RNA-RISC bundles clinging to the mRNA blocks ribosome movement and stops protein synthesis.

Micro-RNAs can also inhibit protein synthesis by destroying mRNA before it can reach ribosomes. In this process, micro-RNA-RISC attaches to the mRNA by base pairing, and then, RISC cleaves the target mRNA. As enzymes rapidly degrade the cleaved mRNA to bits, micro-RNA-RISC seeks its next mRNA target.

Genomes contain regions of duplicated genes. Alterations in gene duplicates can transform a normal gene into a gene that cannot produce a functional protein. A nonfunctional gene is called a pseudogene ("false gene"). For years, scientists considered pseudogenes as junk DNA. However, in 2010, researchers discovered that pseudogenes can play a role in modifying gene expression.

WHEN SMOKE GETS IN YOUR GENES

According to the Centers for Disease Control and Prevention, cigarette smoking accounts for about 443,000 deaths each year in the United States. To put this number into perspective, the agency stresses that tobacco use causes more deaths than the combined number of deaths attributed to human immunodeficiency virus (HIV), illegal drug use, alcohol use, car injuries, suicides, and murders. Cigarette smoke increases the risk of many types of cancer, heart disease, lung disease, and various other health disorders. More than 4,000 compounds can be found in cigarette smoke. Following inhalation of cigarette smoke, these chemicals enter the bloodstream and are distributed to many tissues where they alter gene expression.

Researchers at the Southwest Foundation for Biomedical Research (San Antonio, Texas) analyzed RNA transcripts in white blood cells taken from 1,240 people, including 297 current smokers. Their results show that cigarette smoking affects the expression of at least 323 genes. Studying how cigarette smoke changes gene expression, the researchers suggest, may provide the key to understanding the origins of the adverse health effects attributable to smoking.

The 2010 study focused on a gene called PTEN, which has a counterpart pseudogene dubbed PTENP1. A cell can make mRNA from the PTENP1 pseudogene, but the mRNA has a flaw: PTENP1 mRNA cannot produce a protein. PTENP1 mRNA is not useless, however; the pseudogene's mRNA acts as a decoy for micro-RNAs. Cells that have increased amounts of PTENP1 mRNA (the decoy for micro-RNAs) also have high levels of PTEN protein. Decreasing the amount of PTENP1 mRNA resulted in lower levels of PTEN protein, because micro-RNAs effectively targeted genuine PTEN mRNA.

A cell uses a variety of other tactics to control gene expression. For instance, cells can inactivate certain types of mRNA and store the mRNA until the cell requires the encoded protein. The rate of mRNA translation may be adjusted according to the needs of a cell. An mRNA may be altered to affect its stability. If an mRNA has a long life, then it can be used for translation a number of times. A cell can also control the stability of a protein. When multiple copies of a small molecule called ubiquitin are attached to a protein, that protein is marked for destruction.

THE MANY CONTROL POINTS OF GENE EXPRESSION

A cell uses many strategies to adjust its collection of proteins:

- *Transcriptional control*: Certain proteins turn on or turn off transcription by binding with segments of DNA. In eukaryotes, transcription is further regulated by altering histones and methylating DNA.

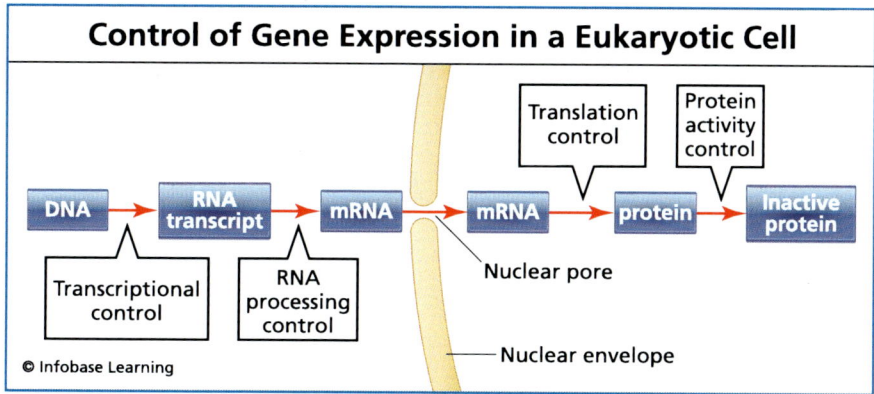

FIGURE 3.3 Eukaryotic cells adjust their collection of proteins and, thus, gene expression.

- *Post-transcriptional control*: Alternative splicing results in the synthesis of different proteins from the same gene. RNA interference blocks protein synthesis by targeting mRNAs. RNA produced from pseudogenes may protect functional mRNAs by diverting RNA interference mechanisms. Cells further regulate protein levels by storing inactivated mRNAs, modifying the rate of translation and stability of mRNA, and by controlling the lifespan of a protein.

Cellular Metabolism

Cellular metabolism is the collection of processes that manages a cell's energy and chemicals. "Energy" is the ability to produce a change, and energy exists as kinetic energy or potential energy. Kinetic energy is the energy of motion or use, whereas potential energy is stored energy. Potential energy can be converted into kinetic energy (and vice versa). For example, the potential energy stored in a coiled spring is converted into kinetic energy when the spring is released. Gasoline has potential energy that is used to provide the kinetic energy that moves a car. The potential energy of gasoline is a type of energy called **chemical energy**. Chemical energy is potential energy stored in chemical bonds; the energy is released in a chemical reaction. Multicellular organisms routinely convert this type of potential energy into kinetic energy. Movement of muscle contraction, for example, is fueled by chemical energy.

For a chemical reaction to occur, chemical bonds must be destabilized. An initial amount of energy, the activation energy, must be supplied to stress a bond before it breaks. Activation energy can be supplied with the addition of heat. Chemical reactions driven by heat would create temperatures much higher than a cell could tolerate. This is where the proteins called enzymes play a vital role. Enzymes are catalysts that accelerate a chemical reaction by reducing the amount of required activation energy. As a result, enzymes ensure that a chemical reaction is more likely to take place at temperatures compatible with life.

41

An enzyme can accelerate a chemical reaction by altering the shape of the substrate, or starting chemical. In this method of action, a substrate binds with an enzyme, causing both to slightly change shape. The substrate is twisted into a form that strains its chemical bonds, so that they break. The substrate forms new chemical bonds and becomes the product of the reaction, which is then released from the enzyme.

The metabolic processes of a cell are organized into pathways of a series of chemical reactions. In a pathway, one enzyme catalyzes a reaction to transform a substrate into a product, which then becomes a substrate for the next enzyme in the sequence.

Enzyme 1   Enzyme 2   Enzyme 3

Substrate 1 $\rightarrow$ Product 1 $\rightarrow$ Product 2 $\rightarrow$ Product 3 $\rightarrow$ (etc.)

Typically, the enzymes of a metabolic pathway are located in a particular part of a cell. Pathway enzymes may even be physically linked to one another.

A cell has two types of metabolic pathways: **catabolic** and **anabolic**. In catabolic pathways, complex molecules are broken down into simpler chemicals. The reverse occurs in anabolic, or "biosynthetic," pathways: Simple chemicals are used to synthesize complex molecules. By breaking down complex molecules, catabolic pathways provide simple chemicals that are the raw materials for anabolic pathways. Catabolic pathways also provide chemical energy required to sustain the life of a cell and to fuel the energy demands of anabolic pathways.

Energy released by a catabolic pathway must be stored until it is needed. Typically, the energy is stored as chemical bond energy in activated carrier molecules. Activated carrier molecules are like batteries that diffuse through a cell to sites where energy is required for biosynthesis and other cell activities. One of the most important and abundant carrier molecules is **ATP (adenosine triphosphate)**, which contains the sugar ribose, the base adenosine, and a chain of three phosphate groups. Energy is stored when a phosphate group (called an inorganic phosphate) is bonded to ADP (adenosine diphosphate) to produce ATP. Energy is released when an inorganic phosphate is split from ATP. This energy-yielding reaction is called a hydrolysis reaction because the ATP molecule is split by the addition of water.

ATP + water $\rightarrow$ ADP + inorganic phosphate + energy

ADP + inorganic phosphate + energy $\rightarrow$ ATP

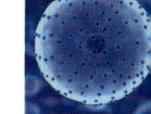

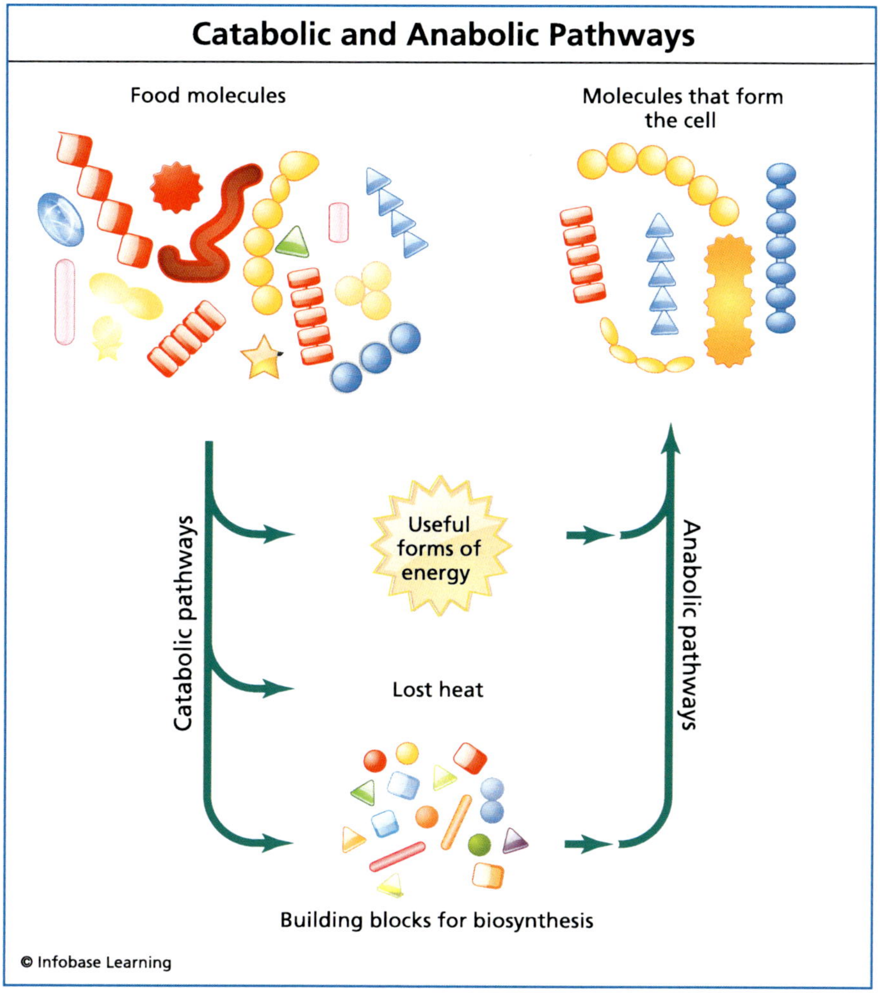

Catabolic and Anabolic Pathways

Food molecules

Molecules that form
the cell

Catabolic pathways

Anabolic pathways

Useful
forms of
energy

Lost heat

Building blocks for biosynthesis

© Infobase Learning

FIGURE 4.1 Most of the energy stored in the chemical bonds of food molecules is released as heat. Thus, the mass of food required by an organism that derives all of its energy from catabolism is much greater than the mass of the molecules that can be produced by anabolism.

ATP is a renewable resource that is continuously recycled in a cell. A functioning muscle cell uses and regenerates 10 million molecules of ATP per second.

Often, the covalent bonds of ATP's phosphate groups are said to be energy-rich bonds. However, the bonds themselves are not high-energy. ATP has high energy relative to ADP. When phosphate is split from ATP, energy is released, because ATP is converted into ADP, which has a lower state of energy. ATP has a high energy state because the three phosphate

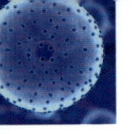

groups are negatively charged and packed close together. The repulsion between negative phosphate groups creates instability. ATP is like a very tightly coiled spring: It is loaded with potential energy.

ATP shuttles energy between catabolic and anabolic pathways. As catabolic pathways generate energy, energy is stored by producing ATP from ADP. Energy needed by anabolic pathways is provided by the release of energy during the conversion of ATP to ADP. Catabolic pathways are like processes that generate money that can be used to pay for the biosynthesis of anabolic pathways.

PRODUCING ATP

Cellular respiration is a key catabolic pathway. Cells use one of two forms of respiration. Aerobic respiration occurs in the presence of oxygen; this process consumes oxygen molecules. A less efficient process, anaerobic respiration, occurs in the absence of oxygen. Many prokaryotic cells and most eukaryotic cells can perform aerobic respiration, whereas some prokaryotes use anaerobic respiration.

Aerobic Respiration

In aerobic respiration, glucose and other organic molecules, such as breakdown products of fats and proteins, are converted to water and carbon dioxide. Along the way, respiration yields chemical energy and small chemicals that the cell uses to construct large molecules in biosynthetic pathways. A cell can convert about 50% of energy stored in a glucose molecule to energy stored in ATP molecules. This is very efficient. An automobile may only change 20% or less of the energy stored in fuel to provide movement; most of the chemical energy is lost as heat. The following overview tracks the catabolism of glucose.

Aerobic respiration takes place in a series of three steps:

1. *Glycolysis:* In the cytosol, a six-carbon glucose molecule is split into two three-carbon molecules of pyruvate and four molecules of ATP are formed. Since glycolysis requires two ATP molecules, the process yields a net production of two ATPs. Two NADH (nicotinamide adenine dinucleotide) molecules also are produced. NADH is another type of carrier molecule. Whereas ATP has high-energy phosphates, NADH has high-energy electrons. Glycolysis does not require oxygen for its chemical reactions.

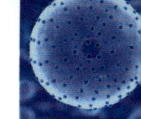

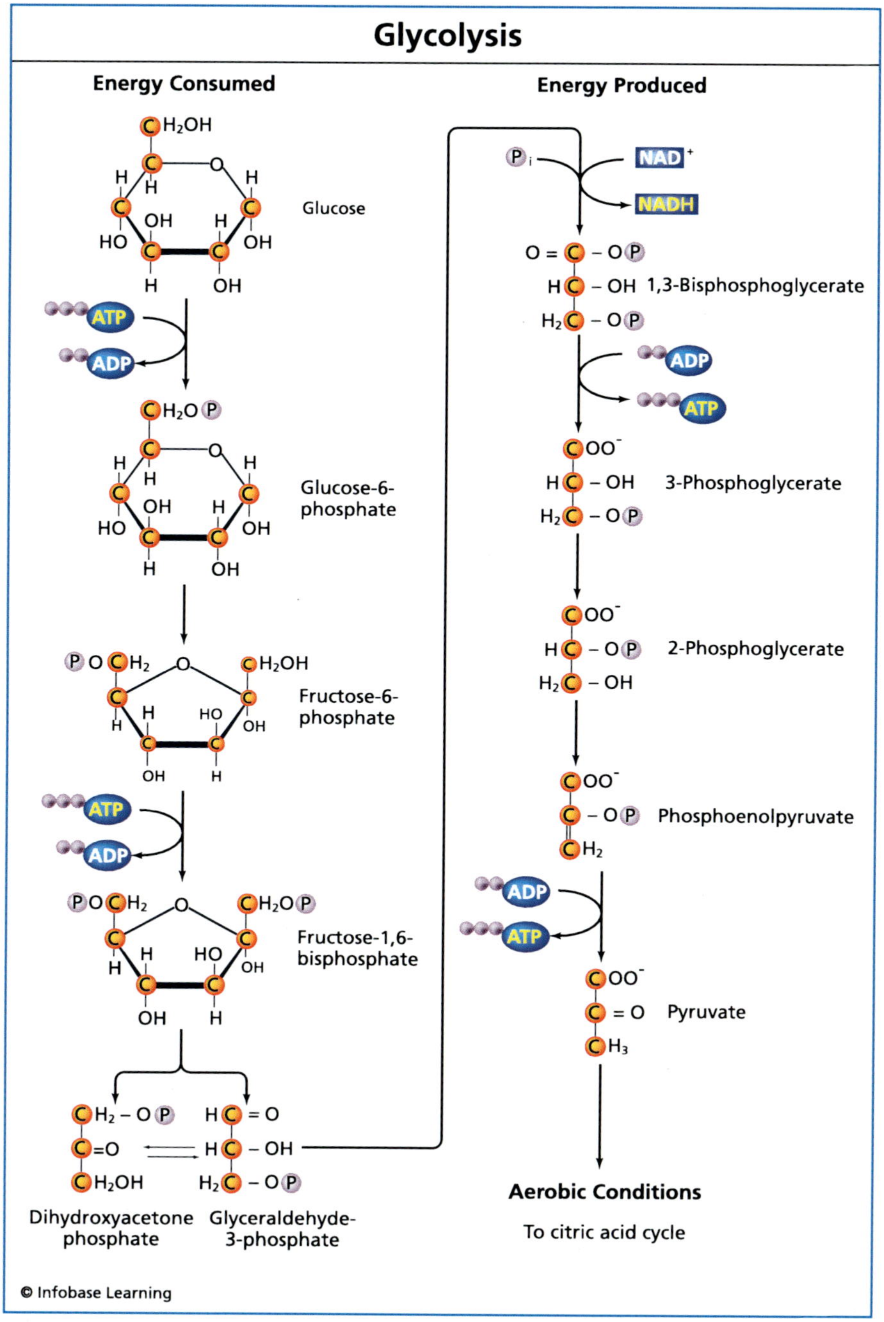

Glycolysis

Energy Consumed

Glucose

Glucose-6-phosphate

Fructose-6-phosphate

Fructose-1,6-bisphosphate

Dihydroxyacetone phosphate   Glyceraldehyde-3-phosphate

Energy Produced

P_i NAD$^+$ NADH

$O = C - O\,P$
$HC - OH$ 1,3-Bisphosphoglycerate
$H_2C - O\,P$

ADP   ATP

COO^-
$HC - OH$ 3-Phosphoglycerate
$H_2C - O\,P$

COO^-
$HC - O\,P$ 2-Phosphoglycerate
$H_2C - OH$

COO^-
$C - O\,P$ Phosphoenolpyruvate
CH_2

ADP   ATP

COO^-
$C = O$ Pyruvate
CH_3

Aerobic Conditions

To citric acid cycle

© Infobase Learning

FIGURE 4.2 During glycolysis, a six-carbon molecule of glucose is broken down into two three-carbon molecules of pyruvate. This generates a net of two molecules of adenosine triphosphate (ATP).

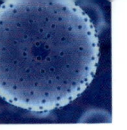

2. ***Citric acid cycle (or Kreb's cycle):*** Each pyruvate molecule is broken down into carbon dioxide and water, generating one ATP and three molecules of NADH. Another type of energy carrier with high-energy electrons, $FADH_2$ (flavin adenine dinucleotide) is also formed. The citric acid cycle reactions require oxygen, and they occur in the cytosol of prokaryotes and within mitochondria of eukaryotic cells.

3. ***Oxidative phosphorylation:*** In this step, electrons are passed from NADH and $FADH_2$ to a series of molecules called electron carriers. As electrons move from one electron carrier to the next, their energy is used to produce ATP from ADP. Three molecules of ATP are formed for each NADH, which contributes two electrons to electron carriers. $FADH_2$ generates two ATPs. At the last step, a molecule of oxygen accepts electrons to form water. The electron transport chains are located in the plasma membrane of prokaryotes and on the inner membranes of mitochondria in eukaryotic cells.

If ATP were money, then oxidative phosphorylation would be the real moneymaker of the entire respiration process. Whereas glycolysis and the citric acid cycle produce 2 ATPs each from a glucose molecule, oxidative phosphorylation generates 32 to 34 ATPs.

Producing ATP Without Oxygen

Certain prokaryotes use anaerobic respiration to generate ATP. The process includes the use of an electron transport chain. Some bacteria that dwell in tidal flats and marshes, for example, use an electron transport chain to generate ATP and hydrogen sulfide (H_2S), rather than ATP and water (H_2O). Anaerobic respiration produces lower amounts of chemical energy than aerobic respiration. Yet, the energy yield is higher than that obtained from fermentation.

Fermentation produces chemical energy in the absence of oxygen and without use of an electron transport chain. Two common types of fermentation are lactic acid fermentation and alcohol fermentation. Certain fungi and bacteria perform lactic acid fermentation, in which pyruvate molecules generated by glycolysis are converted into lactic acid. This type of fermentation also takes place in animal muscles under exertion, a condition which creates an oxygen-poor environment for muscle cells. Anyone who feels sore muscles after a workout knows about the effect of lactic acid. A practical use of lactic acid fermentation is the production of yogurt

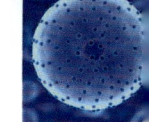

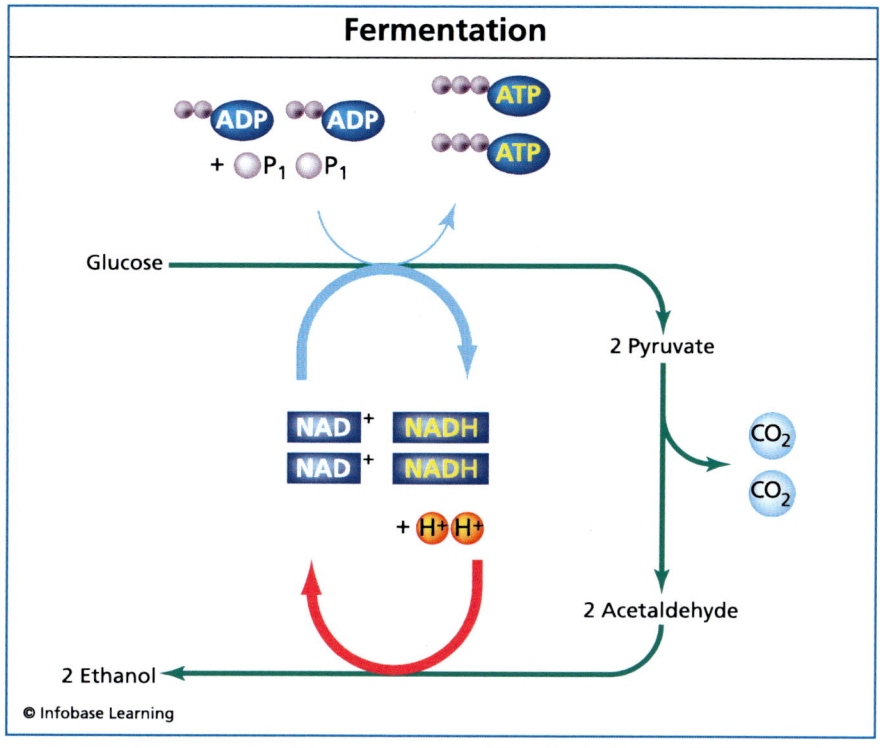

Fermentation

FIGURE 4.3 Under anaerobic conditions, yeast convert glucose into pyruvic acid via the glycolysis pathways, and then move a step farther, converting pyruvic acid into ethanol.

and cheese. Yeast and certain bacteria perform alcohol fermentation after glycolysis. In this process, pyruvate molecules are converted into carbon dioxide and ethanol. Humans take advantage of alcohol fermentation to make bread, wine, and beer.

Compared with respiration in the presence of oxygen, fermentation is an inefficient method for producing ATP. Aerobic respiration can produce about 36 molecules of ATP; fermentation yields only two ATPs.

REGULATION OF METABOLISM

Recall how the bacterial trp and lac operons are regulated. The trp operon is repressed when trp binds with the trp operon repressor protein, whereas the binding of lactose with the lac operon repressor results in the stimulation of the lac operon. This difference is logical because the trp operon encodes enzymes required to produce trp, and a cell should turn off syn-

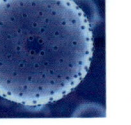
MITOCHONDRIAL DISEASES

Human mitochondria contain a circular, 16,500-base pair molecule of DNA that includes about 37 genes. Thirteen mitochondrial genes encode enzymes that are involved in oxidative phosphorylation. The other mitochondrial genes encode RNA molecules required for synthesis of the enzymes: 22 types of transfer RNAs and 2 types of ribosomal RNAs. Nuclear DNA encodes other proteins required for a mitochondrion to function. These proteins are produced in the cytoplasm, and they are transported to mitochondria. Unlike the genes of nuclear DNA, mitochondrial genes of mammals are usually inherited only from the mother. During fertilization, a sperm's nuclear DNA enters the egg and combines with the egg's nuclear DNA. Most of the sperm's mitochondria are left outside the egg. Although small numbers of sperm mitochondria enter the egg, paternal mitochondrial DNA is usually destroyed. Consequently, the fertilized egg typically contains only maternal mitochondrial DNA.

Mitochondrial DNA can carry mutations that are maternally inherited. A gene **mutation** is a change in the nucleotide sequence of a DNA molecule. A mutation in a gene encoding a protein may result in an alteration in the amino acid sequence of the encoded protein or a mutation in a gene can prevent synthesis of the encoded protein. Scientists have identified more than 250 mitochondrial DNA mutations associated with a disease. Many mitochondrial diseases share a common feature: The DNA mutations hinder the ability of mitochondria to produce sources of chemical energy. Defects in mitochondria particularly affect muscle cells and nerve cells, which require great amounts of energy to function. Hearing and vision loss, seizures, and muscle weakness are common symptoms experienced by those who have a mitochondrial disease. An effective therapy or cure is unavailable for patients with a mitochondrial disease. Researchers are focusing on developing treatments to minimize a patient's symptoms and prevent disease complications.

thesis of the biosynthetic enzymes if plenty of trp is available. On the other hand, the lac operon encodes proteins needed to accumulate and break down lactose. To conserve energy, a cell should synthesize the proteins

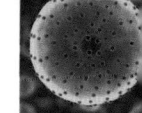

only if lactose becomes available. A wide variety of mechanisms control metabolic pathways. One tactic is similar to the function of repressors in trp and lac operons: A small molecule binds with a protein and changes the protein's shape to alter the protein's function. In the case of metabolic regulation, the protein is an enzyme. A second method of control requires a modification of an enzyme by the covalent attachment of a chemical group, such as a phosphate. A third control strategy targets the regulation of enzyme synthesis.

Regulation of metabolism is complex. Consider just one aspect of the regulation of a key enzyme of glycolysis. As discussed above, a cell obtains chemical energy from ATP by splitting it into ADP and inorganic phosphate. Sometimes, two inorganic phosphates are cleaved from ATP to form AMP (adenosine monophosphate), or AMP is formed by splitting an inorganic phosphate from ADP. An enzyme called PFK (phosphofructokinase) catalyzes an early step in glycolysis, and its activity is affected by the amounts of ATP and AMP. ATP inhibits the activity of PFK, whereas AMP stimulates PFK enzyme activity. This makes sense. If a cell has plenty of ATP, then it has no need to run glycolysis and the other respiration reactions at full throttle. However, a buildup of AMP means that the cell is running out of chemical energy. So, AMP stimulates PFK to ramp up glycolysis for the generation of more ATP.

In addition to mechanisms for regulating respiration and other catabolic pathways, a cell regulates anabolic (biosynthetic) pathways. Again, it is useful to focus on one aspect of glucose metabolism—in this case, the production of glucose, which is called gluconeogenesis. Glycolysis and gluconeogenesis are regulated within a cell to ensure that one pathway is active while the other is rather inactive. A cell could not survive if both pathways were highly active at the same time: Glycolysis would degrade glucose, which would be replaced by gluconeogenesis with a net loss of chemical energy. In business terms, running both glucose pathways at once would bankrupt the cell.

Since cells are very efficient, catabolism and anabolism of glucose are regulated. Take another look at glycolysis, in which two chemical reactions—represented by two arrows—convert glucose to fructose 6-phosphate. The PFK enzyme then catalyzes the following reaction:

$$\text{Glucose} \rightarrow \rightarrow \text{fructose 6-phosphate} + \text{ATP} \xrightarrow{\text{PFK}} \text{fructose 1,6-bisphosphate} + \text{ADP}$$

In gluconeogenesis, the reverse reaction is catalyzed by the enzyme FBPase (fructose 1,6-bisphosphatase):

$$\text{Fructose 1,6-bisphosphate + water} \xrightarrow{\text{FBPase}} \text{fructose 6-phosphate + phosphate} \rightarrow \rightarrow \text{glucose}$$

AMP, which signals low chemical energy, stimulates PFK of glycolysis, but AMP inhibits FBPase of gluconeogenesis. In this way, a cell with low stores of chemical energy will generate more ATP via glycolysis, and it will not invest in the expenditure of its low stores of ATP by producing more glucose via gluconeogenesis.

PHOTOSYNTHESIS

Algae, certain bacteria, and plants perform photosynthesis, a process that uses energy from sunlight to create organic (carbon-containing) molecules from atmospheric carbon dioxide. The general formula for the photosynthetic reaction is:

$$\text{Light energy + water + carbon dioxide} \rightarrow \text{sugar + oxygen}$$

Plant cells contain membrane-bound chloroplasts, which have pigments that capture light. Chlorophyll is a light-absorbing photosynthetic pigment. Large amounts of chlorophyll can be found in chloroplasts of leaf cells. Leaves—the major site of photosynthesis in most plants—appear green because chlorophyll absorbs blue and red light, leaving green light to be reflected from the leaf. Within a chloroplast, a membrane forms a set of flattened sacs called thylakoids, which contain electron transport chains and pigments that capture light. In the first stage of photosynthesis, chloroplasts convert energy from light into the chemical energy of ATP and the activated carrier molecule NADPH (nicotinamide adenine dinucleotide phosphate). As electrons move through the electron transport chain, water is split, producing oxygen. During the next stage of photosynthesis, ATP and NADPH fuel reactions that produce sucrose and other carbohydrates from carbon dioxide. These reactions start in the stroma, a region within chloroplasts and outside the thylakoids, and continue in the cytosol. The sugars can be degraded for chemical energy or used to synthesize molecules needed by the cell.

Photosynthesis has profoundly affected life on Earth. The process is responsible for oxygen in the atmosphere and the production of organic

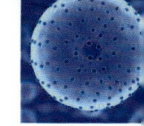

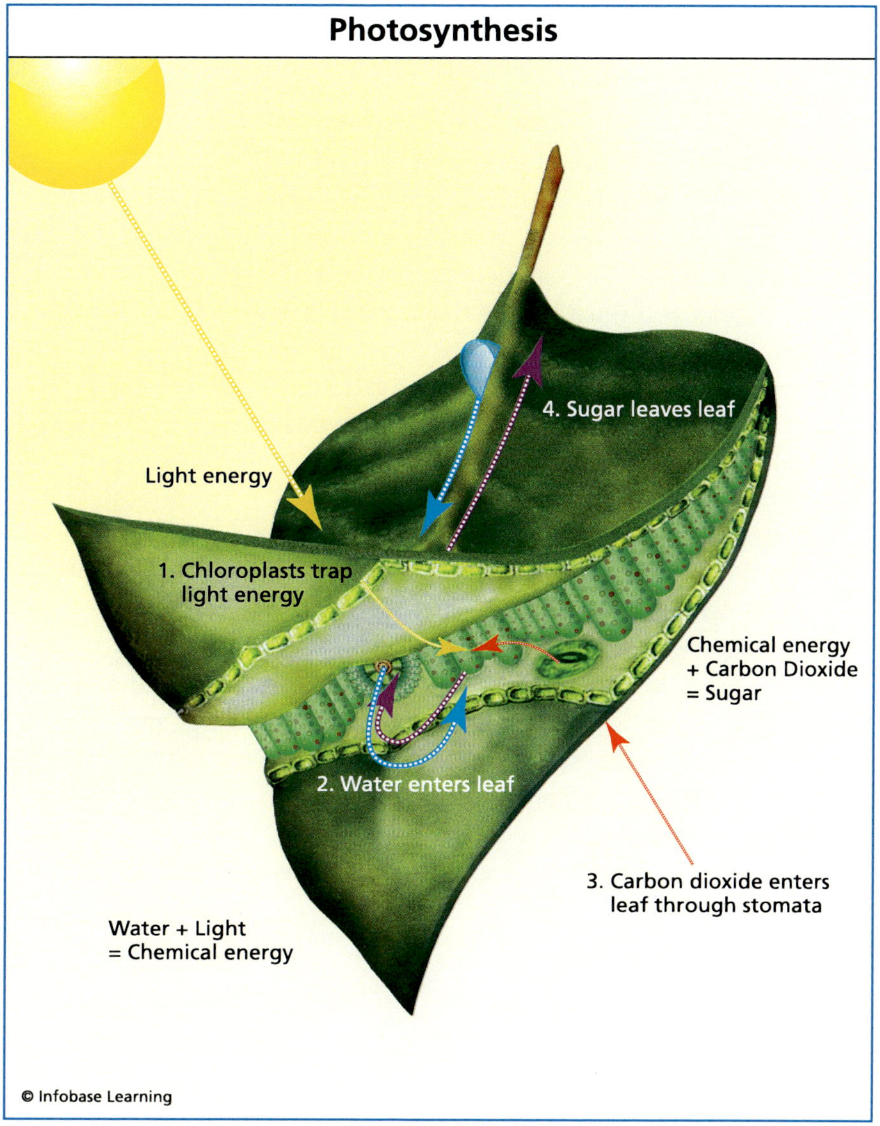

Photosynthesis

Light energy

4. Sugar leaves leaf

1. Chloroplasts trap light energy

Chemical energy + Carbon Dioxide = Sugar

2. Water enters leaf

3. Carbon dioxide enters leaf through stomata

Water + Light = Chemical energy

© Infobase Learning

FIGURE 4.4 In photosynthesis, carbon dioxide is converted into organic compounds, including sugars, using light energy.

molecules that serve as nutrition for humans and other organisms that cannot perform photosynthesis. The oxygen generated as a by-product of photosynthesis is used by cells to perform aerobic respiration, the process that ultimately forms carbon dioxide from oxygen. Photosynthesis and cellular respiration form a cycle:

GLYCOGEN STORAGE DISEASES

Animals store glucose in the form of the glucose polymer known as glycogen. When necessary, enzymes degrade glycogen into glucose or glucose-6-phosphate in a process called glycogenolysis. In a reverse process called glycogenesis, enzymes synthesize glycogen from glucose. Although glycogen can be found in many types of human tissue, it is liver cells and muscle cells that contain the greatest amounts of glycogen. The two cell types use glycogen for different purposes. Glycogen stored in muscle cells provides carbohydrates, which the cells can use to generate ATP to power contraction. Liver glycogen serves to store glucose until the sugar is required to be released from the cells to maintain blood glucose levels.

Glycogen storage diseases are a group of inherited diseases in which glycogen breakdown is defective or cells contain abnormal forms of glycogen. In 1929, Edgar von Gierke was the first to describe a glycogen storage disease, which was named after him. (The disease is characterized, in part, by a low level of glucose in the blood and an enlarged liver with an excess of glycogen.) Twenty-three years later, Carl and Gerty Cori discovered the cause of von Gierke disease: Liver cells lacking glucose 6-phosphatase, the enzyme that is necessary to convert glucose 6-phosphate into glucose. Without the enzyme, the breakdown of glycogen stops with glucose 6-phosphate, a charged molecule that cannot pass through the liver cell plasma membrane. Since liver cells cannot complete the conversion of glycogen to glucose, blood glucose levels decrease below normal. Von Gierke disease is also caused by a mutation in the gene encoding a protein that transports glucose 6-phosphate to glucose 6-phosphatase.

Von Gierke disease is the most common glycogen storage disease. Researchers have characterized at least eight other types of glycogen storage diseases. Although treatment depends on the particular form of glycogen storage disease, management of the disease usually requires people who have the condition to regulate their consumption of carbohydrates.

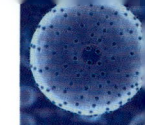

Photosynthesis: carbon dioxide and water → oxygen and organic molecules

Aerobic respiration: oxygen and organic molecules → ATP, carbon dioxide, and water

INCOMING SIGNALS ALERT CELLS TO ALTER METABOLISM

Cellular metabolism can be summarized as two coupled processes: (1) the generation and expenditure of chemical energy, and (2) the breakdown and construction of complex molecules. The survival of a cell depends upon the finely tuned regulation of the many metabolic pathways. Single-celled organisms modify metabolism in response to changes in their environment. The cells of multicellular organisms control metabolic pathways to meet the needs of the entire organism. In both cases, data about conditions outside a cell must be transferred to appropriate locations within a cell. This is the topic of cell communication.

5

Cell Communication

A CLOSER LOOK AT THE PLASMA MEMBRANE OF ANIMAL CELLS

A plasma membrane serves as a wrapper that holds together the organelles and cytosol of a cell. Yet the membrane is not an inert covering; it regulates the flow of molecules into and out of the cell. In scientific terms, a plasma membrane is selectively permeable.

The outer membrane of a cell contains lipids, proteins, and carbohydrates. The most common type of membrane lipid is phospholipid. Recall that a lipid is typically hydrophobic and avoids interacting with water. A phospholipid is a type of lipid that has both hydrophobic and hydrophilic regions. The molecule can be pictured as having a hydrophilic head and two hydrophobic tails. Most plasma membranes have two layers of phospholipids that are mirror images. In the outside layer, the hydrophilic heads of phospholipids face the watery exterior of the cell, and the phospholipid hydrophobic tails are buried inside the membrane. Phospholipids of the inside layer position their hydrophobic tails inside the membrane and place their hydrophilic heads toward the watery interior of the cell. A membrane is a fluid structure, and most lipids move laterally within a membrane.

A plasma membrane also contains proteins, which can also shift their position within the membrane. Scientists have identified more than 50 different types of plasma membrane proteins. These proteins are peripheral

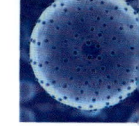

Plasma Membrane

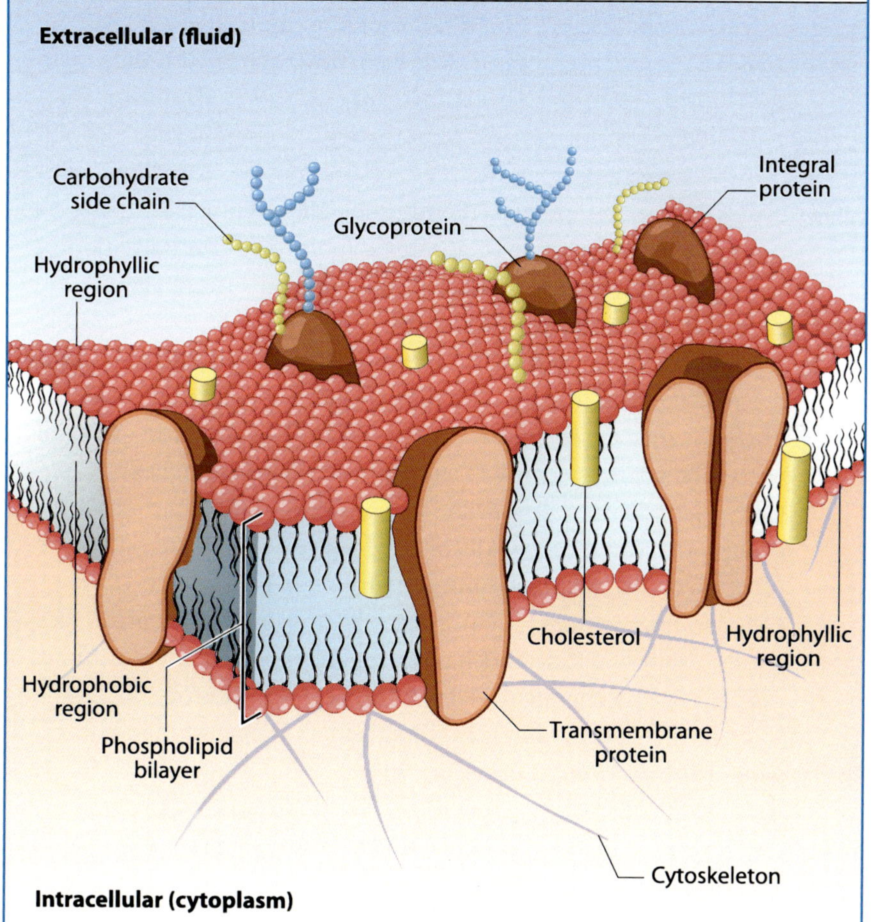

Extracellular (fluid)

Carbohydrate side chain

Glycoprotein

Integral protein

Hydrophyllic region

Cholesterol

Hydrophyllic region

Hydrophobic region

Transmembrane protein

Phospholipid bilayer

Cytoskeleton

Intracellular (cytoplasm)

© Infobase Learning

FIGURE 5.1 A plasma membrane is made up of two phospholipid layers with embedded proteins. It regulates what enters and exits the cell.

proteins or integral proteins. Peripheral proteins extend partly into the hydrophobic region, bobbing along the surface of the fluid membrane. Integral proteins penetrate the hydrophobic region of the phospholipid bilayer. Some integral proteins—which are called transmembrane proteins—extend through the membrane, so that parts of the protein are exposed at both exterior and interior surfaces of the membrane.

Plasma membrane proteins perform functions vital to a cell:

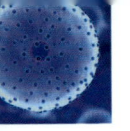

- **Molecule carriers:** Diffusion is a process in which molecules spontaneously move from areas of high concentration to areas of low concentration. Small hydrophobic molecules can diffuse across a plasma membrane. Some transmembrane proteins form hydrophilic channels that allow diffusion of hydrophilic molecules that cannot readily pass through a membrane. Certain transport proteins act as ATP-energized pumps, propelling molecules across the plasma membrane. These transport proteins require energy, because they shuttle molecules from a region of low concentration to a region of high concentration, against the molecule's concentration gradient.
- **Cell connectors:** Membrane proteins of one cell can join with membrane proteins of a neighboring cell, especially in tissue.
- **Maintenance of cell shape:** Proteins of the cytoskeleton can bind with membrane proteins to support the shape of the cell.
- **Cell identifiers:** Glycoproteins—proteins with carbohydrate groups—face the cell's exterior and serve as identifiers that are recognized by plasma membrane proteins of other cells.
- **Cell communication:** Membrane proteins called **receptors** face the exterior of a cell and have a region that binds with a particular type of chemical. After binding, the receptor signals the interior of the cell, which can result in a change of metabolism, gene transcription, or other activity.

CELLS USE SIGNALING MOLECULES AND RECEPTORS TO COMMUNICATE

Cell signaling with receptor proteins occurs in three stages. In the first stage, a small molecule binds with a receptor protein and causes the receptor protein to react. For example, the binding of the small molecule can stimulate the receptor to alter its shape. The small molecule is often called a signaling molecule or first messenger. During the second stage, the altered receptor protein passes a signal on to another cell component. This is called **signal transduction**, because the initial signal—the binding of a first messenger with its receptor—is converted into another type of signal, such as the attachment of a phosphate group to a protein. A number of proteins may be involved in the signal transduction stage, with each protein relaying the signal to another protein in a signaling pathway. In the third stage, a signal causes a cellular response, such as an increase in the transcription of a particular gene.

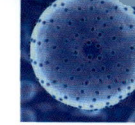

Receptor binding → signal transduction → cellular response

As the name "first messenger" implies, cell signaling can involve other messengers. In the signal transduction process, the binding of the first messenger with its receptor may result in the synthesis or release of many second messenger molecules. An example of a second messenger is cyclic AMP, a modified form of AMP. Second messenger molecules diffuse through the cell and spread the signal. Depending upon the particular signaling pathway, a signal will be passed from one signaling protein to another. Some signaling pathways, for example, require signaling proteins that gain or lose a phosphate group, an alteration that activates or inactivates the signaling protein.

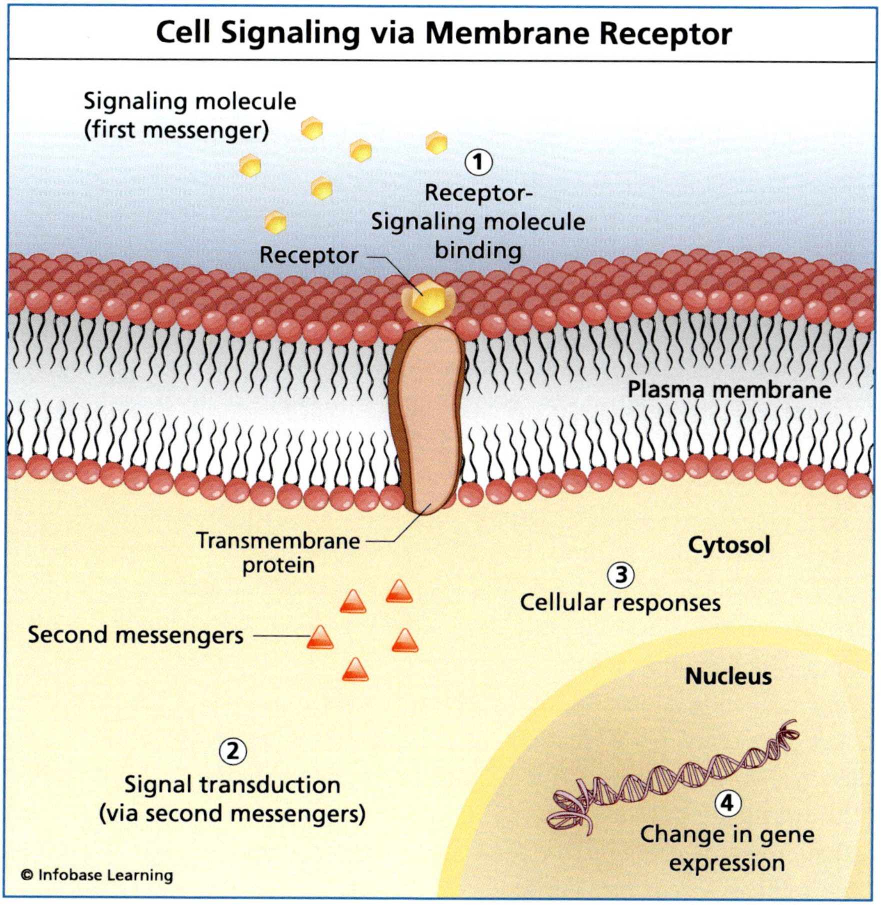

Cell Signaling via Membrane Receptor

Signaling molecule
(first messenger)

① Receptor-
Signaling molecule
binding

Receptor

Plasma membrane

Transmembrane
protein

Cytosol

③
Cellular responses

Second messengers

Nucleus

②
Signal transduction
(via second messengers)

④
Change in gene
expression

© Infobase Learning

FIGURE 5.2 Some signaling molecules bind with receptor proteins located in plasma membranes. This interaction can alter cellular activities.

First messengers are typically hydrophilic and cannot pass through the hydrophobic core of a plasma membrane. Instead, they bind with receptor proteins, which convey a signal to the cell's interior. Some small, hydrophobic signaling molecules can slide through a plasma membrane and into a cell. For example, thyroid hormones and steroid hormones diffuse across plasma membranes. Once inside a cell, these hormones bind with intracellular receptors. The binding activates the receptors, which then affect the transcription of certain genes. In some cases, activated intracellular receptors stimulate transcription; in other cases, activated receptors block gene transcription.

BACTERIA SENSE WHEN THEY HAVE A QUORUM

Multicellular organisms rely upon cell communication to coordinate the functions of various tissues and cells. Single-celled organisms rely upon cell communication to respond to changes in their environment. Many bacteria react to chemicals secreted by other bacteria living nearby. A rise in the concentration of these chemicals indicates increasing numbers of cells. A bacterium's ability to sense rising population density is called quorum sensing, and it enables a group of bacteria to coordinate many activities, including reproduction.

Thanks to quorum sensing, many disease-causing bacteria produce toxic substances when a bacterial population is sufficiently large to ensure a successful attack against the defenses of an infected host. *Pseudomonas aeruginosa*, for example, is a type of quorum-sensing bacteria that infects humans and secretes disease-causing toxic factors. Cell-to-cell signaling systems enable *P. aeruginosa* to coordinate the synthesis of hundreds of toxins, depending upon the density of the bacterial population. The bacteria's quorum-sensing system requires proteins that bind with signaling molecules. Two genes encode the binding proteins and also play a role in the creation of a biofilm. A biofilm, which connects layers of bacteria, protects infecting bacteria against defenses of the host. Scientists are devising ways to treat *P. aeruginosa* infections with drugs that target the proteins that are encoded by the two genes and the bacteria's quorum-sensing system.

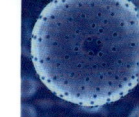

A cell of a multicellular organism responds to signaling molecules released near the cell or released from another part of the body. One type of local signaling occurs when cells secrete signaling molecules that only affect one or more neighboring cells. When a nerve cell is stimulated, for example, the nerve cell releases chemicals that diffuse to a target cell, such as another nerve cell or a muscle cell, and stimulate a change in the target cell. Endocrine glands signal over long distances by secreting signaling molecules—hormones—that circulate in the bloodstream or other fluids.

CONSEQUENCES OF CELL SIGNALING IN MAMMALIAN GLUCOSE METABOLISM

Glucose is a vital source of energy for mammals. To ensure the availability of glucose, cells store the sugar in the form of glycogen, a glucose polymer. Muscle cells degrade glycogen into glucose, which is then used to generate ATP to power contraction. Liver cells break down glycogen to form glucose, which is released into the bloodstream. Many processes regulate the use and storage of glucose.

Glucagon Stimulates Glucose Release into the Bloodstream

The protein hormone glucagon helps to maintain the level of glucose in the blood. Cells of the pancreas secrete glucagon if the amount of glucose in the blood falls below a certain level. Glucagon travels through the bloodstream and stimulates the release of glucose from liver cells. How glucagon achieves this activity provides an example of a signaling pathway.

1. Glucagon binds with glucagon receptors in the plasma membrane of liver cells.
2. Glucagon binding with its receptor activates a membrane protein called the G protein.
3. Activated G protein activates the enzyme adenylate cyclase, a membrane protein that faces the interior of the cell.
4. Activated adenylate cyclase converts ATP into cyclic AMP (a second messenger).
5. Increased levels of cyclic AMP in the cytosol activate protein kinase A.
6. Activated protein kinase A catalyzes the attachment of phosphate groups to phosphorylase kinase and glycogen synthase.

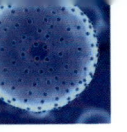

This activates phosphorylase kinase and deactivates glycogen synthase, the enzyme that converts glucose to glycogen.

7. Activated phosphorylase kinase activates glycogen phosphorylase by attaching a phosphate group.

8. Activated glycogen phosphorylase stimulates glycogen breakdown into glucose.

In this pathway, the signal is modified, or transduced, as follows:

Receptor binding → generation of cyclic AMP → phosphate group attachment

The overall effect of the signaling pathway is to increase the release of glucose from the liver cell by decreasing the conversion of glucose to glycogen and increasing the breakdown of glycogen to glucose.

Each step of the pathway amplifies the initial signal. The binding of one glucagon molecule to its receptor may result in the production of thousands of cyclic AMP molecules, which eventually leads to the activation of hundreds of thousands of glycogen phosphorylase enzymes. Activation of the glucagon signaling pathway could rapidly deplete a liver cell's stores of glycogen. However, many processes also modulate the amplified signals by silencing them. For example, the enzyme phosphodiesterase converts cyclic AMP into AMP, ending the mission of the second messenger. Enzymes called phosphatases modify phosphorylase kinase and glycogen phosphorylase by removing their phosphate groups. The combination of signal amplification and signal reduction fine tune a cell's metabolism.

Insulin Stimulates Glucose Uptake and Glucose Stimulates Insulin Secretion

Pancreas cells also secrete insulin, another protein hormone. Insulin binds to receptor proteins located in the plasma membrane, and this initiates changes that allow glucose to travel from the bloodstream to the cell's interior, where glucose is used to generate chemical energy or glucose is stored as glycogen.

The insulin receptor is a transmembrane protein. The extracellular region of the insulin receptor binds insulin, resulting in a change in the shape of the intracellular region of the receptor, which stimulates the receptor to attach phosphate groups to itself. The activated insulin receptor stimulates signaling pathways involving proteins and second messengers. Stimulation of the insulin signaling pathway results in the movement of glucose transporter proteins from the cytoplasm to the plasma membrane. An increased number of glucose transporter proteins in the membrane leads to increased glucose uptake from blood. Insulin binding to its receptor also promotes the synthesis of glycogen from glucose.

Insulin acts through a receptor-mediated signaling pathway to stimulate cells to remove excess glucose from blood. What provokes pancreas

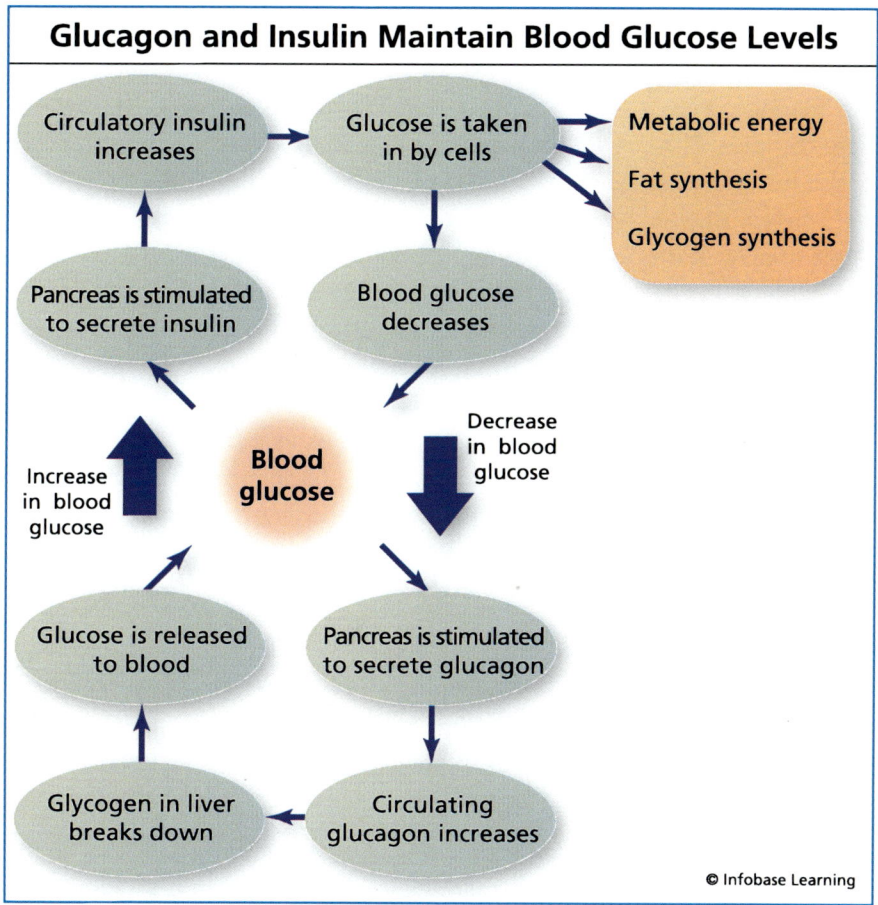

Glucagon and Insulin Maintain Blood Glucose Levels

Circulatory insulin increases

Glucose is taken in by cells

Metabolic energy

Fat synthesis

Glycogen synthesis

Pancreas is stimulated to secrete insulin

Blood glucose decreases

Increase in blood glucose

Blood glucose

Decrease in blood glucose

Glucose is released to blood

Pancreas is stimulated to secrete glucagon

Glycogen in liver breaks down

Circulating glucagon increases

© Infobase Learning

FIGURE 5.3 This insulin and glucagon regulation model shows that as blood glucose levels rise, insulin is secreted in the pancreas and circulated throughout the body; glucose is then taken in by cells and blood glucose levels decrease. When blood glucose levels decrease, the pancreas secretes glucagon to break down glycogen in the liver and releases glucose into the blood.

cells to release insulin? The answer is high levels of glucose in the bloodstream. Glucose stimulates insulin secretion, although it does not use a receptor-based signaling system. As concentrations of blood glucose

PLANT CELLS KEEP IN TOUCH

As complex multicellular organisms, plants must organize many types of processes within their cells. A plant must coordinate functions in its leaves, roots, and stems. Changes in temperature and light conditions affect a plant's flowering, fruiting, and growth. A plant uses hormones to manage many processes of its cells. The actions of the hormones are complex. Although plants appear to use only six types of hormones, each hormone can generate numerous effects. Furthermore, the action of one type of plant hormone can be modified by the presence of another type of plant hormone, as well as by environmental factors.

One plant hormone, ethylene, promotes fruit ripening and other normal plant activities; ethylene also signals the presence of dangerous conditions, such as flooding, drought, and injury. Ethylene receptors are transmembrane proteins. When ethylene receptors are free of ethylene, the receptors add phosphate groups to proteins as a signal transduction mechanism. The signals inactivate gene regulatory proteins that stimulate the transcription of certain genes. The ethylene signaling system is switched on as a default condition. After ethylene binds with its receptor, the ethylene signal system is switched off, allowing gene transcription to proceed.

Receptor without ethylene: inactivates regulatory proteins that would stimulate transcription

Ethylene binds receptor: regulatory proteins are now free to stimulate transcription

Plant cells directly communicate with neighboring cells through plasmodesmata. These small channels connect cytoplasm of adjoining plant cells. Nucleic acids, proteins, and other molecules diffuse through plasmodesmata. These communication channels also enable a virus infection to spread throughout a plant. Plant viruses can stimulate the channels to open wide enough to allow them to travel from cytoplasm to cytoplasm of neighboring cells.

climb, glucose transporter proteins shuttle the sugar across the plasma membrane. Inside the cell, glucose is metabolized to produce ATP. As more ADP is converted to ATP, the ATP : ADP ratio rises, causing certain protein channels in the plasma membrane to close. When they are open, these protein channels allow positively charged potassium ions to exit the cell. With the channels shut, potassium ions accumulate, resulting in an increase in the positive charge within the cell. Other membrane proteins react to the increased positive charge by allowing calcium ions to rush into the cell. A pancreas cell can store at least 10,000 granules of insulin waiting for release. The rapid influx of calcium is the signal that triggers secretion of the stored insulin.

Increased intracellular glucose

↓

Glucose is metabolized to produce ATP from ADP

↓

Increased ATP : ADP ratio causes shutdown of potassium ion channel

↓

Potassium ions accumulate in cell, increasing the positive charge

↓

Elevated positive charge triggers calcium ion channels to open

↓

Calcium ions stimulate insulin release

Glucose provokes another response in pancreas cells: The sugar stimulates transcription of the insulin gene. In human cells, glucose appears to activate a signaling pathway that results in the recruitment of proteins required for transcription and the relaxation of chromatin structure to allow transcription proteins to access the insulin gene promoter.

SAME SIGNAL, DIFFERENT RESULT

Signaling molecules transmit information to a cell from its environment. Many signaling molecules perform their function by binding with receptors in the plasma membrane. Some signaling molecules pass through the plasma membrane and bind with intracellular receptors. In both cases,

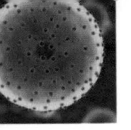

receptor binding is converted into other types of signals that flow through a signal pathway and alter cell functions.

Different types of cells in a multicellular organism react differently to the same extracellular signaling molecules. The response of a cell to a signaling molecule depends upon the cell's set of proteins, including receptor proteins, signal pathway proteins, and proteins that perform a cell's ultimate response to a signaling molecule. Even if two types of cells have receptors that bind the same signaling molecule, the two cell types may not respond in the same way. For example, certain nerve cells release a chemical called acetylcholine, which binds with a plasma membrane receptor of a target cell. Depending upon the signal pathway of a target cell, acetylcholine receptor binding causes a target cell to contract (such as skeletal muscle cells), to decrease contraction (heart muscle cells), or increase secretion (salivary gland cells).

6

Cell Communication Successes

Cells modify various functions in response to extracellular signaling molecules that bind with membrane receptors. Cell communication, however, is not limited to signaling molecules. Cells also communicate with each other by direct contact. In one form of direct contact, molecules pass from the interior of one cell to the interior of a neighboring cell. Another type of direct contact relies upon interactions between membrane proteins of two cells.

CELLS GET ORGANIZED

Single-celled organisms are the oldest type of life on Earth. Over time, multicellular life forms arose, with each increasing level of complexity representing greater cooperation between cells. Scientists categorize cell cooperation into five types.

The **protoplasmic level of organization** is represented by single-celled organisms. For this type of organism, all life-sustaining functions occur within a cell, where organelles perform specialized functions.

A **cellular level of organization** is characterized by a division of labor among differentiated cells that live together. Differentiated cells are cells that perform specific functions. At the cellular level of organization, differentiated cells do not form tissues, which are a collection of similar cells that perform a common function. The sea-dwelling sponge is an example of an organism with a cellular level of organization. Although a sponge lacks true tissues, a sponge does have collections of specialized cells that

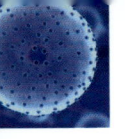

are loosely organized in a jellylike matrix. One type of specialized cell are collar cells, which have flagella and line the inside of a sponge. Sponges collect food particles from water by producing currents with the beating flagella of collar cells.

Organisms, such as jellyfish, with the **cell-tissue level of organization** have similar cells assembled into tissues. In the center of their sac-shaped bodies, jellyfish possess a gastrovascular cavity, which functions in digestion and circulation of nutrients. Tentacles surround the opening into the gastrovascular cavity and push prey into the cavity. Once food is inside the gastrovascular cavity, an inner layer of tissues secretes digestive juices to break down the food. The resulting nutrients are circulated among cells within the cavity.

In the **tissue-organ level of organization**, tissues assemble into organs. An organ has a more specialized function than a tissue and is usually composed of more than one type of tissue. Consider a type of flatworm called a planarian, which possesses muscles, nerves, and digestive organs. A typical planarian has an arrow-shaped head with sensory organs to detect food or danger. Planarians secrete mucous to entangle their prey. After wrapping itself around its trapped food, a planarian extrudes an organ called a pharynx through its mouth. Although the pharynx is similar to a muscular throat, it is located around the middle of the planarian's body. The pharynx ejects digestive fluids on its food and sucks small pieces of partially digested food into a gastrovascular cavity for further digestion.

In the **organ-system level of organization**, organs work together to perform a function to create an organ system, such as a respiratory system. For example, mammals have an immune system consisting of a group of organs that work together to defend the body against foreign invaders and poisonous substances. Bone marrow, the spleen, tonsils, and other tissues play a role in the immune response.

Functions of Junctions

The five categories of cell cooperation can be summarized as follows:

- Cells survive individually.
- Specialized cells work together.
- Specialized cells form tissues.
- Tissues form organs.
- Organs work together to create organ systems.

REGENERATING TISSUES

According to the U.S. Organ Procurement and Transplantation Network, there were more than 108,000 people in the United States waiting for organ transplants in mid-2010. Patients needed livers, lungs, hearts, kidneys, and other organs. In the field of tissue engineering, researchers devise methods for growing human tissue and organs in the laboratory. Regenerated tissue may be transplanted into patients to replace failed tissues.

The rat provides a popular model for testing tissue regeneration technology. For example, in 2010 a team led by Yale University scientists reported that they had rebuilt rat lungs. The researchers removed lungs from dead rats and treated the lungs with a mild detergent to burst the lung cells. The detergent treatment left behind blood vessels, airway structures, and extracellular matrix. The researchers suspended this remaining material in a glass container and added nutrients and lung cells from rat fetuses. Using the blood vessels and airway structures as scaffolds, the fetal cells formed tissue in the matrix and created the complex structure of a lung within days. The researchers transplanted the new lungs into live rats, and, for several hours, the lungs functioned like normal lungs. Scientists speculate that human lung tissue may be regenerated from extracellular matrix that would be removed from the lungs of a cadaver. However, studies with regenerated human lungs and patients may be decades away.

In 2008, University of Minnesota researchers reported that they had invented a way to produce a beating rat heart in their laboratory. After removing a heart from a dead rat, the scientists treated the heart with detergent to destroy cells. They added fetal rat cells to a mixture of extracellular matrix, blood vessels, and heart valves. In two weeks, the cells had formed a heart that pumped blood and conducted electrical impulses.

In 2009, another science team at the University of Michigan reported that it had regenerated tissue by coaxing the cells into creating more gap junctions. The researchers genetically altered cells to synthesize excess amounts of gap junction proteins. An increased number of the proteins resulted in a higher number of gap junctions and increased communication between cells in the new tissues.

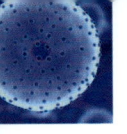

Scientists propose that a vital step in the evolution of multicellular life forms was the development of the ability of cells to contact each other securely and to interact with neighboring cells. Animal cells secrete proteins and carbohydrates that form a network called the **extracellular matrix**, which helps to unite cells in tissues. Several types of junctions between animal cells also strengthen the bond between cells and allow cell communication.

Tight junctions form when the plasma membranes of neighboring cells press against each other and are bound together by proteins. The junctions create continuous seals around cells. A leak-proof sheet of tissue, for example, lines the digestive tract and tight junctions between skin cells enable skin to create a watertight covering.

Anchoring junctions rivet neighboring cells to each other or attach cells to the extracellular matrix. Although anchoring junctions secure adjacent cells to each other, they do not create a tight seal: Molecules can still pass through spaces between the cells.

Gap junctions provide channels that pass from the cytoplasm of one cell to the cytoplasm of a neighboring cell. The channel is composed of an integral membrane protein called connexin. In the plasma membrane, six connexin transmembrane proteins form a ring around a central opening. This structure is called a connexon. An intercellular channel is formed when the connexons of two adjacent cells align with each other. After formation, a gap junction allows the exchange of small molecules, such as ions, amino acids, and various metabolites. The plasmodesmata of plant cells, mentioned in the previous chapter, perform a similar function.

Animal cell gap junctions play a vital role in cell-to-cell communication and integrate the activities of a tissue's cells so that they function as a unit. By sharing metabolites, for example, cells cooperate in metabolic activities. Gap junctions also enable a stimulus to rapidly pass to all cells of a tissue. For instance, gap junctions between heart muscle cells electrically couple cells to enable coordinated muscle contraction in a healthy heart. Human heart diseases characterized by an irregular heartbeat typically involve heart muscle cells with abnormal gap junctions.

THE ANIMAL IMMUNE SYSTEM: DEFENDING THE BODY WITH CELL COMMUNICATION

The immune system is a remarkable organ system that protects the body against billions of bacteria, viruses, parasites, and toxins. Bone marrow, the spleen, tonsils, and other tissues produce and store white blood cells,

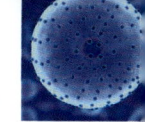

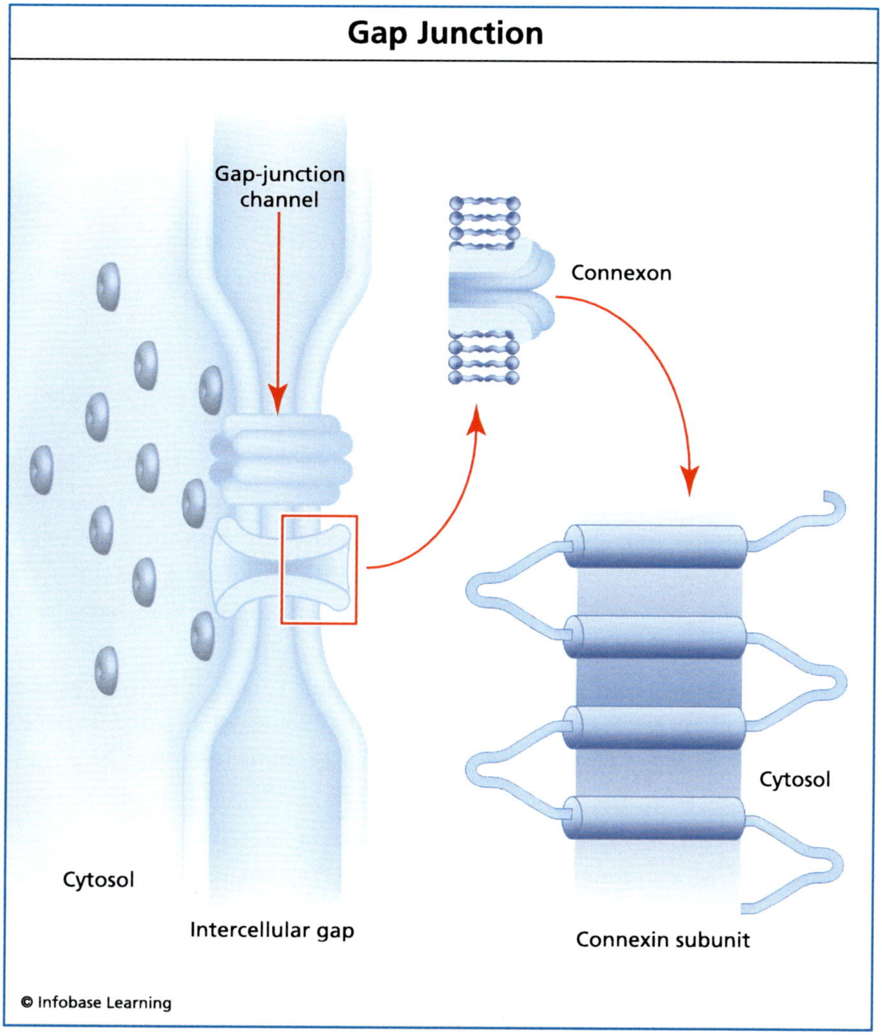

Gap Junction

Gap-junction channel

Connexon

Cytosol

Cytosol

Intercellular gap

Connexin subunit

© Infobase Learning

FIGURE 6.1 Six transmembrane proteins form a ring around a central opening (a connexon). The connexons of two cells form a gap junction.

which are also called lymphocytes. White blood cells attack foreign invaders. This means that immune system cells must distinguish between cells that belong in the body and cells that are foreign to the body. In other words, immune system cells can tell the difference between cells that are "self" and those that are "non-self."

All cells of an animal display plasma membrane proteins that identify the cells as self. Normally, an animal's immune system cells do not attack the body's own cells that carry the self markers. This normal state is known

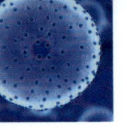

as self-tolerance. When immune system cells detect invading cells, such as bacteria, which lack the self-markers, the immune cells attack the foreign cells. A non-self molecule that can trigger an immune response is called an **antigen**. Many types of foreign substances are antigens, including a whole virus or bacterium, a carbohydrate, a protein, or a part of a protein.

Scientists have identified two types of immune responses: **innate immunity** and **adaptive immunity**. While all animals have innate immunity, only vertebrates—animals with a backbone that encloses a nerve cord—have adaptive immunity.

Innate Immunity

Innate immunity provides a first line of defense against foreign cells, viruses, and toxins. This type of immunity is innate in the sense that an animal is born with these defenses. In effect, they are "hardwired" into the body. Innate immunity is also called nonspecific immunity, because some innate immunity defenses identify something as foreign to the body by recognizing generic signals. For example, common polysaccharides in bacterial cell walls can trigger the innate immune system.

Insects possess multiple components of their innate immune system. An insect body is covered with a hard, but flexible, exoskeleton composed of layers of protein and the polysaccharide known as chitin. The exoskeleton protects an insect against physical hazards and helps to prevent dehydration, while its inside layer provides a surface for the attachment of muscles. The exoskeleton also creates a barrier against many disease-causing bacteria. Invading bacteria bypass the exoskeleton barrier when an insect eats food contaminated with bacteria. However, the invaders can be blocked by a chitin barrier in the intestinal tract. The digestive system also produces lysozymes, enzymes that degrade a component of bacterial cell walls. The insect circulatory system includes a heart that pushes a fluid through short vessels and into spaces that surround tissues and organs. If bacteria slip by the intestinal track defenses and enter the circulatory system, the invaders will encounter insect cells that defend the body. Some of these defensive cells are phagocytes, which are cells that ingest and destroy bacteria. Other defensive cells stimulate the secretion of chemicals that kill bacteria.

The innate immune systems of humans and other vertebrates also have barrier defenses and cellular defenses. Barrier defenses include skin and the cornea of the eye. Cells lining nasal passageways and other parts of the respiratory tract secrete mucus that traps bacteria. Tears and saliva

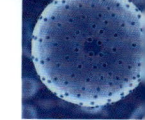

contain lysozymes and other proteins that destroy bacteria. If invading cells and viruses reach the stomach, they face the corrosive effects of acid.

Once inside the bloodstream, bacteria encounter many other defensive tactics of innate immunity. For example, the human body has about 30 proteins that make up the complement system. Inactive complement proteins circulate in the bloodstream until they bind with molecules displayed by many types of bacteria. Binding activates the complement proteins, which results in chemical reactions that destroy the bacteria. Phagocytic cells patrol the bloodstream and respond to certain polysaccharides in bacterial cell walls by engulfing and digesting bacteria.

Natural killer cells, a type of white blood cell, will destroy a potential target cell, depending upon a mixture of receptor signals. Natural killer cells have two types of plasma membrane receptors: activating and inhibiting receptors. When stimulated, activating receptors spur the natural killer cell to kill. Stimulated inhibitory receptors discourage the cell from killing. The balance of signals between these two types of receptors determines whether a natural killer cell will slay a potential target cell. A natural killer cell usually spares the life of a cell that carries self markers. Immune system cells distinguish between self and non-self by recognizing whether a cell has MHC (major histocompatibility complex) proteins in its plasma membrane. Class I MHC proteins are displayed on the surface of most healthy cells and stimulate a natural killer cell's inhibitory receptors. Alternatively, unusual carbohydrates or proteins on the surface of a cell, such as a bacterial cell, stimulate the activating receptors. Virus-infected cells pose a more interesting challenge. A cell infected with a virus can display viral proteins and Class I MHC proteins in its plasma membrane. Such an infected cell stimulates both activating and inhibiting receptors of a natural killer cell. Scientists do not know how a natural killer cell "evaluates" a blend of signals. After taking over a cell, some viruses unwisely shut off expression of Class I MHC molecules. Natural killer cells receive only kill signals from these infected cells.

Adaptive Immunity

If innate immunity defenses fail to stop an invading pathogen, then the vertebrate body responds with adaptive immunity measures. Innate immunity is present at birth, whereas adaptive immunity is acquired and takes time to mount a defense. Adaptive immunity tailors a response to a specific antigen, which may be a protein or carbohydrate that the body recognizes as foreign.

White blood cells that are called B cells play an important role in adaptive immunity. When a B cell finds an antigen that binds with its plasma membrane receptors, the cell begins to change. The B cell reproduces itself many times to create two types of cell populations: plasma cells and B memory cells. The plasma cells synthesize antibodies, which are globular proteins that bind with the same antigen that bound the receptor of the B cell. Plasma cells can secrete their antibodies at a rate of tens of thousands of antibodies per second.

Antibodies serve many functions:

- Antibodies can bind with antigens displayed by an invading cell or a virus-infected cell, and mark the cell for destruction by phagocytes.

FORENSIC SCIENTISTS TRACK POLLEN

During the late nineteenth century, a radical idea percolated in Europe that would revolutionize criminal investigations: the use of scientific methods to investigate and solve crimes. In 1910, Dr. Edmond Locard founded the first police crime laboratory in Lyon, France. Locard also promoted a principle that became the cornerstone of forensic science. He said that a cross-transfer of physical evidence occurs when any person comes into contact with another person or an object. For example, a person who visits a particular place will deposit hairs, dirt from shoes, and other material, and that person will leave with dust, dirt, and other substances from the area.

One type of material that easily passes from a person to a place—and vice versa—is pollen. Protected within a durable shell, a grain of pollen can travel great distances in the wind and in moving water. Pollen also adheres to hair and clothing. Forensic palynology is the study of pollen and other micro-plant remains for legal purposes. Pollen is useful in criminal investigations because it resists degradation, occurs widely in the environment, and has distinctive shapes that enable identification of particular plant types. Experts in this field regularly offer evidence in criminal trials in Australia, New Zealand, and the United Kingdom, and, to a lesser extent, in the United States. Forensic palynology has been used to connect a suspect to the scene of a crime, connect an item left at

- Antibodies can bind with a toxin that carries the antigen and neutralize the toxin.
- Antibody binding to cells that display the antigen can activate the proteins of the complement system, which play their part in destroying the cell.
- Antibodies can coat viruses that carry the matching antigen and prevent the viruses from infecting cells.
- Antibodies can bind groups of viruses and decrease the number of infectious viruses in the blood.

A T cell is another type of white blood cell that plays an important part in adaptive immunity. The plasma membrane of a T cell contains antibody-like receptors that bind with antigens on the surface of virus-

a crime scene with a suspect, reconstruct the travel of illegal drugs and other evidence, and provide data about the geographic origin of an item.

Police in New Zealand used pollen identification expertise in a case about a break-in. One night, a woman awoke to find two male intruders in her bedroom. In their hurry to leave the house, one of the intruders brushed against a flowering *Hypericum* bush that grew near the back door. The police arrested a suspect and sent his clothes to a lab for forensic examination. Analysis revealed significant amounts of *Hypericum* pollen that was identical in shape, size, color, and age to pollen removed from the *Hypericum* bush at the victim's house. Although this evidence was circumstantial, the pollen traces helped to build a case against the suspect.

Pollen experts also apply their skills in lawsuits. In 1989, a pilot and his wife died when their twin-engine plane crashed in New Mexico. The couple's children hired attorneys whose investigation of the plane's wreckage revealed the existence of plant material inside the fuel line of an engine. They then sued the airplane manufacturer on the grounds that the crash occurred as a result of the engine sucking in vegetation that clogged the fuel line. Palynologists for the defense explained that the vegetation contained pollen grains that were intact and showed no signs of damage from a blazing crash. Furthermore, the palynologists traced the pollen to plants that grew in the yard where the plane wreckage had been stored. A fuel line blocked with plant material had not caused the crash, the experts concluded.

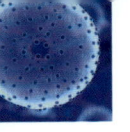

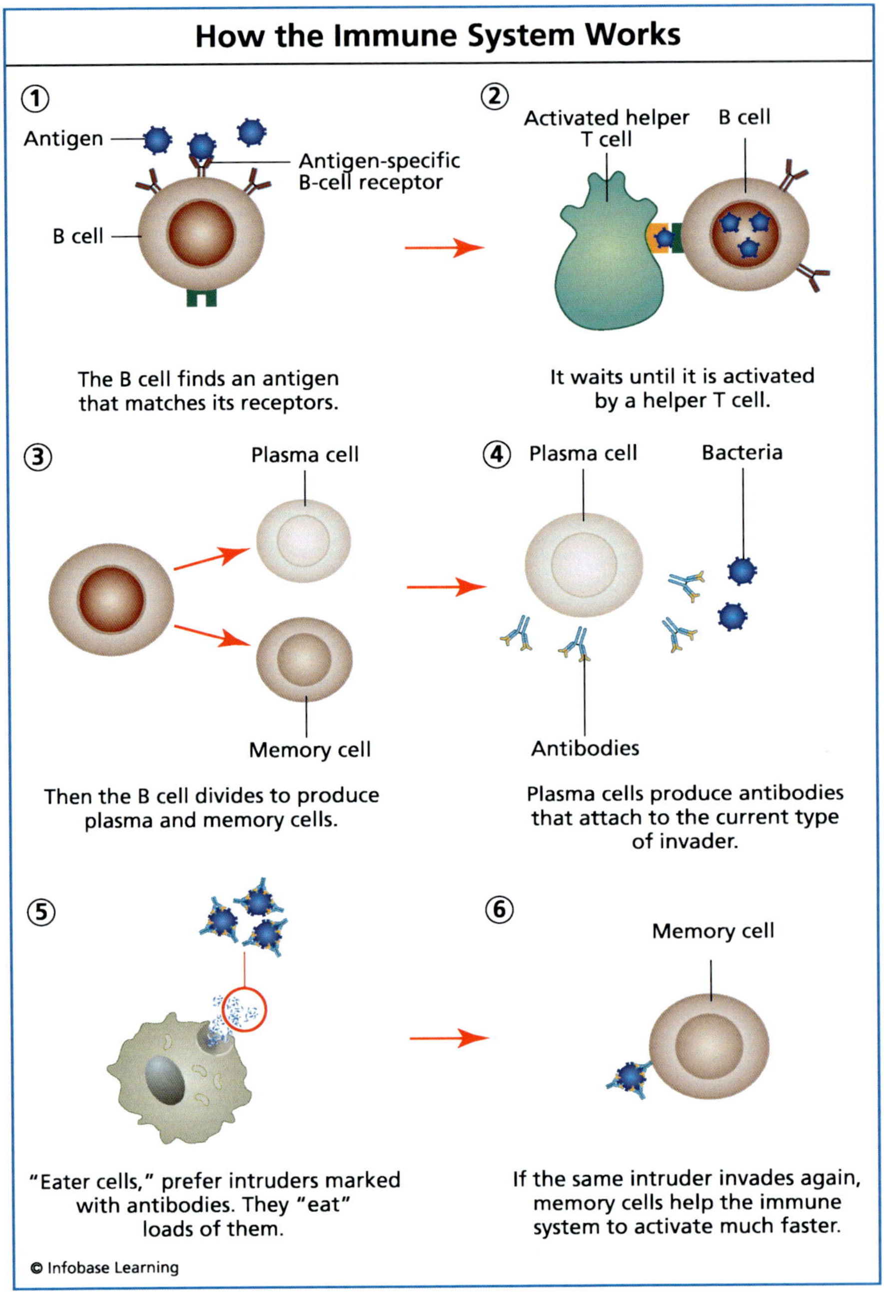

How the Immune System Works

①

Antigen

Antigen-specific B-cell receptor

B cell

The B cell finds an antigen that matches its receptors.

②

Activated helper T cell   B cell

It waits until it is activated by a helper T cell.

③

Plasma cell

Memory cell

Then the B cell divides to produce plasma and memory cells.

④

Plasma cell   Bacteria

Antibodies

Plasma cells produce antibodies that attach to the current type of invader.

⑤

"Eater cells," prefer intruders marked with antibodies. They "eat" loads of them.

⑥

Memory cell

If the same intruder invades again, memory cells help the immune system to activate much faster.

© Infobase Learning

FIGURE 6.2 The immune system protects against disease by identifying and killing pathogens and tumor cells that intrude the body.

infected cells or bind with abnormal antigens on cancer cells. T cells support an immune response in many ways. Helper T cells stimulate B cells to synthesize antibodies, attract phagocytes to an area infested with

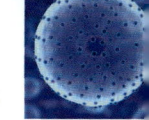

invading cells, and activate killer T cells. Killer T cells attack cells that display foreign molecules on their surface. For example, killer T cells recognize virus antigens in the plasma membrane of infected cells and kill the infected cells.

Every B cell or T cell carries many plasma membrane receptor proteins that bind one particular antigen. This enables the adaptive immune system to mount a specific, targeted response. The adaptive immune system also protects the body against a broad range of disease-causing cells and toxins. The immune system achieves this extensive protection with a collection of millions of white blood cells equipped with receptor proteins that bind different antigens.

As mentioned above, activated B cells produce memory cells. T cells also produce memory cells. Memory cells have long life spans and patrol the body. If a foreign cell with antigens that match the receptors of memory cells invades the body, the memory cells help the immune system to mount a much quicker response. Memory cells enable this greater efficiency for several reasons. When the body first encounters a foreign cell, the immune system may include about one B cell or T cell in a million that can recognize the foreign cell's antigens. After a foreign cell stimulates the adaptive immune system, about one in a thousand B or T cells can recognize the antigens. Furthermore, memory B cells and memory T cells are easier to activate than the original, inexperienced B cells and T cells. The immune system is geared up for a second attack. In this way, immunological memory offers protection against infectious diseases from a prior infection.

LINES OF COMMUNICATION

Without the different types of cell communication, multicellular organisms could not exist. Cell communication that takes place via extracellular signaling molecules enables cells to adjust their various activities according to the needs of the body. Gap junctions allow communication between neighboring cells that transforms a group of cells into a functional unit. Immune system cells protect the body using information received from plasma membrane receptors that interact with other membrane proteins or foreign molecules in bacterial cell walls. Cell communication also signals a cell when it is time for the cell to reproduce or time for the cell to die.

7

Cell Reproduction and Cell Death

A cell divides to create two new cells. The original cell is called the parent cell and the new cells are called daughter cells. Eukaryotic cell division can achieve one of two objectives: reproduction of identical cells or generation of genetic diversity. **Mitosis** is an asexual reproduction process that typically produces identical daughter cells, whereas **meiosis** produces non-identical cells—egg cells and sperm cells—for sexual reproduction. In sexual reproduction, genes from two individuals combine into new arrangements. The ability to shuffle genes among members of a species was a significant development during the evolution of life on Earth. Sexual reproduction enables a species to create genetic variation, which gives rise to new combinations of traits among its members. New assortments of characteristics can become critical for the survival of a species in a changing environment.

CELL DIVISION THAT REPRODUCES CELLS

Binary Fission

Prokaryotes reproduce by **binary fission** in which a parent cell divides into two daughter cells that are usually identical to the original cell. Before binary fission, a cell synthesizes a copy of its single DNA molecule. Recall that DNA is a polymer of nucleotides and that each nucleotide has a base that sticks out from DNA's sugar-phosphate backbone. The four bases of a DNA molecule are adenine (A), cytosine (C), guanine (G), and thymine

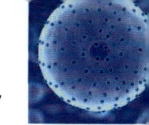

Replication of DNA

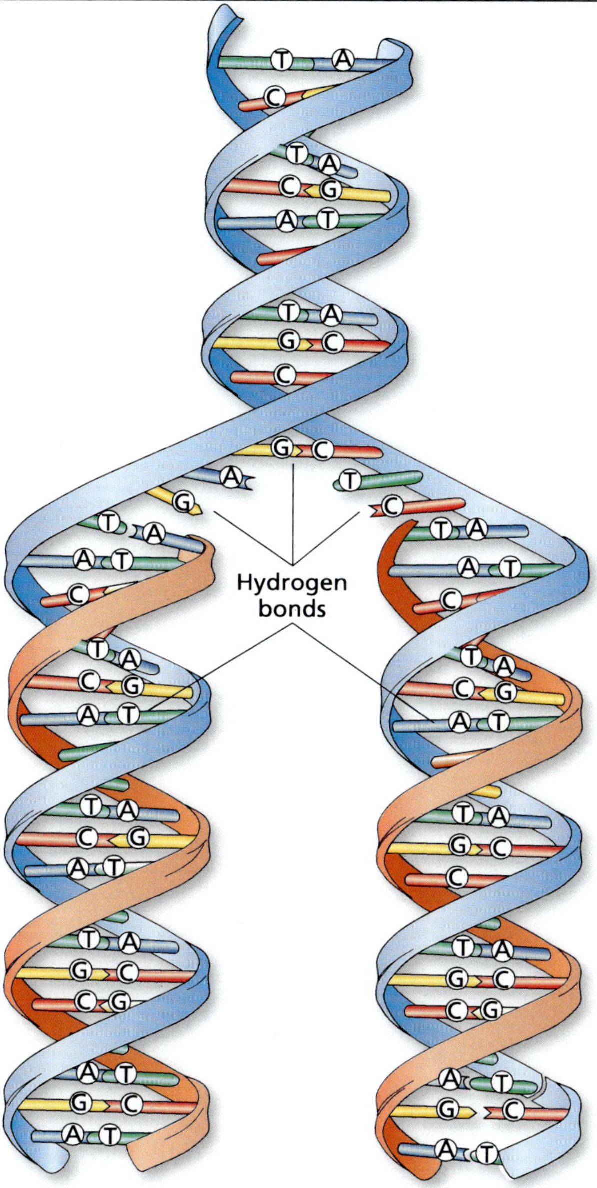

Hydrogen
bonds

© Infobase Learning

FIGURE 7.1 DNA's power to duplicate itself is explained by its structure.
When a cell prepares to divide, the hydrogen bonds between the bases
dissolve and the DNA molecule splits along its length like a zipper
unzipping. Each half then attracts nucleotides, forming the same pairs of
bases that had existed before. The result is two identical DNA molecules.

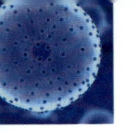

(T). Two strands of DNA form a double-stranded helix, because an A on one DNA strand pairs with a T on the other DNA strand, and a G on one DNA strand pairs with a C on the other DNA strand. That is, bases of two different DNA strands bind together and form a base pair.

The formation of base pairs is critical for nucleic acid synthesis. As discussed previously, an RNA molecule is synthesized during transcription by the enzyme RNA polymerase as the enzyme travels along a DNA molecule, combining nucleotides to synthesize the RNA polymer. The enzyme incorporates nucleotides into the RNA molecule that can form base pairs with the nucleotides in the DNA strand, which is called the template DNA strand. Two nucleotide sequences that can form a series of base pairs are said to be complementary nucleotide sequences.

A cell synthesizes a copy of its DNA molecule using a process that is similar to transcription. During DNA replication, an enzyme unwinds the DNA double helix to allow other enzymes access to the two DNA molecules. Enzymes called DNA polymerases travel along a DNA molecule and combine nucleotides to produce a new DNA polymer. A DNA polymerase incorporates nucleotides into the new DNA polymer that are complementary to the nucleotides in the template DNA strand. Following DNA synthesis, a prokaryotic cell has two double-stranded DNA molecules, and each of these molecules consists of one original DNA strand and one newly synthesized DNA strand. DNA replication in prokaryotes—as well as in eukaryotes—is called semiconservative replication, because one strand of the original DNA molecule is conserved in each new DNA double helix.

After a prokaryotic cell duplicates its DNA molecule, each DNA molecule is attached to a different part of the cell membrane. When the cell divides by pulling itself apart, the two DNA molecules are distributed between the two halves. In this way, each daughter cell typically contains an identical DNA molecule.

The Eukaryotic Cell Cycle and Mitosis

Eukaryotes have two types of cells: **somatic cells** and **gametes**. In a multicellular eukaryote, the majority of cells are somatic cells. The exceptions are egg cells and sperm cells, which are also called gametes. The eukaryotic **cell cycle** of somatic cells starts with the division of a parent cell into new daughter cells. The cycle continues until a new daughter cell completes its own process of division and produces its own two daughter cells. The cell cycle consists of four phases:

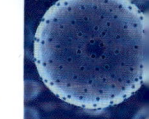

- *Gap 1 (G1) phase:* The G1 phase is the interval, or gap, between the creation of a new cell and DNA synthesis.
- *Synthesis (S) phase:* During the S phase, the cell synthesizes histone proteins and DNA for copies of each chromosome. At the completion of the S phase, every DNA molecule has a duplicate.
- *Gap 2 (G2) phase:* The G2 phase is the interval between DNA synthesis and mitosis.
- *Mitosis (M) phase:* The M phase begins with the division of the nucleus by mitosis and ends when the parent cell splits into two cells by cytokinesis.

The completion of cytokinesis leaves two small daughter cells with depleted stores of ATP. During the G1 phase, the new cells generate ATP and increase in size. When a cell has sufficient chemical energy, DNA synthesis begins. Once again, the cell must replenish its stores of ATP after the S phase and during the G2 phase. The cell needs energy for mitosis. When they are not engaging in mitosis, cells also duplicate their cytoplasmic organelles.

Cells regulate progression through phases of the cell cycle with checkpoints. Three well-known checkpoints occur between the G1 and S phases, between the G2 and M phases, and during the M phase. The cycle stops at a checkpoint if the previous step has not achieved certain goals, or if certain conditions are not favorable for cell division. For example, a checkpoint that operates in the G1 phase ensures that sufficient nutrients are available before the cell proceeds to the S phase. The presence of damaged DNA after the S phase causes an arrest of the cell cycle in the G2 phase. The cell must repair any DNA damage before proceeding to mitosis. Another checkpoint helps to ensure that daughter cells receive equal amounts of DNA. This checkpoint is triggered by the presence of unreplicated DNA after the S phase. The cell cycle stops in the G2 phase until all DNA is replicated. After all DNA has been copied, the cycle can proceed to mitosis. A checkpoint during mitosis ensures that chromosomes are correctly distributed between the two halves of the cell.

Eukaryotic cells differ in the amount of time required for a cell cycle. Some human cells, such as mature nerve cells, do not divide at all. Cells in the kidney, liver, and other tissues divide when necessary to replace injured cells and cells that have died. Other rapidly dividing cells have a 24-hour cycle with a G1 phase of about 11 hours, an S phase that lasts about 8 hours, a G2 phase of 4 hours, and an M phase of only 1 hour.

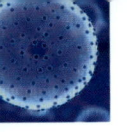

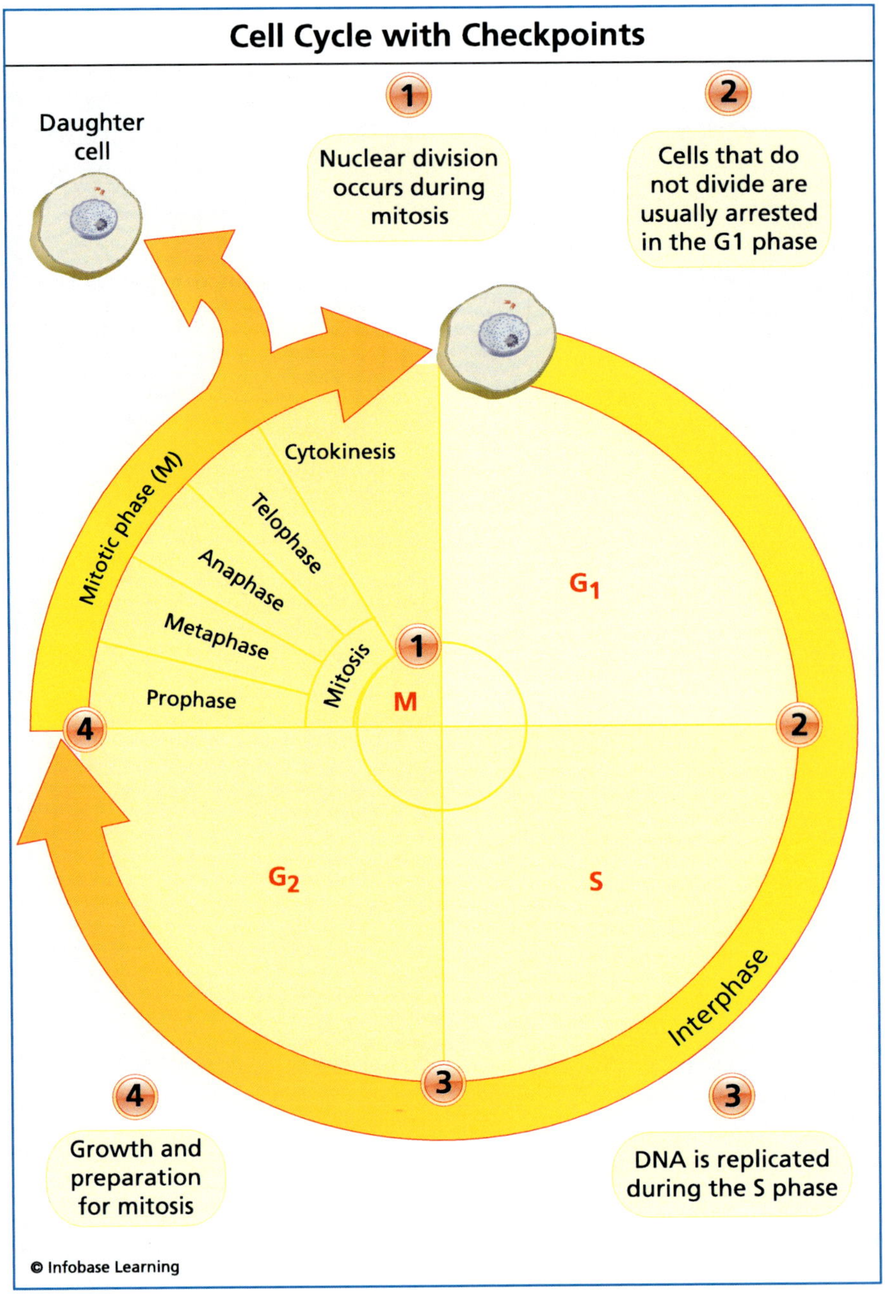

Cell Cycle with Checkpoints

1
Nuclear division occurs during mitosis

2
Cells that do not divide are usually arrested in the G1 phase

Daughter cell

Cytokinesis

Mitotic phase (M)

Telophase

Anaphase

Metaphase

Prophase

Mitosis

1

M

G₁

4

2

G₂

S

Interphase

4
Growth and preparation for mitosis

3

3
DNA is replicated during the S phase

© Infobase Learning

FIGURE 7.2 The cell cycle is a series of events that take place inside a cell, thereby leading to cell division and duplication. G1 and G2 are growth stages, during which the cell matures. The chromosomes replicate during the S (synthesis) phase. During the M (mitotic) phase, the cell undergoes the stages of mitosis, as well as cytokinesis, in which the two daughter cells are formed.

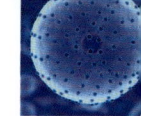

Although the M phase of rapidly dividing cells is short, much happens during this part of the cell cycle.

When a cell is ready to divide, mitosis proceeds through four stages:

1. **Prophase:** Long molecules of DNA compress into densely packed chromosomes. The membrane that surrounds the nucleus—the nuclear envelope—breaks down. Duplicate chromosomes are called sister chromatids. Sister chromatids contain identical DNA molecules. Under the microscope, sister

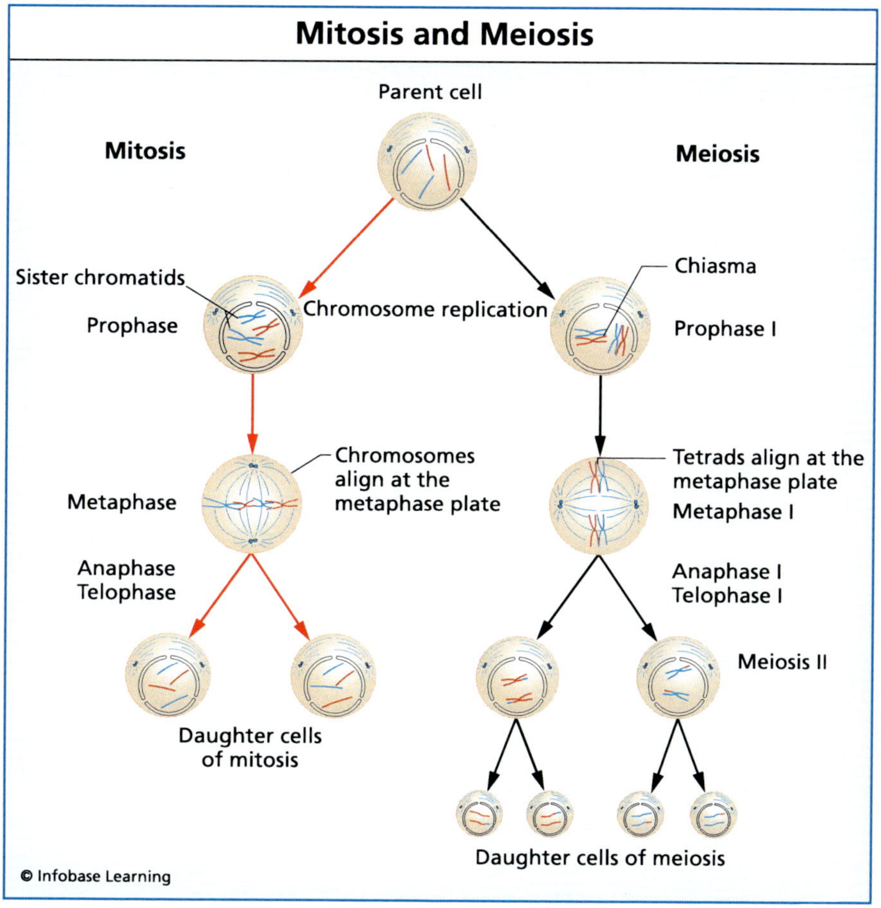

Mitosis and Meiosis

Parent cell

Mitosis

Meiosis

Sister chromatids

Prophase

Chromosome replication

Chiasma

Prophase I

Chromosomes align at the metaphase plate

Metaphase

Tetrads align at the metaphase plate

Metaphase I

Anaphase
Telophase

Anaphase I
Telophase I

Meiosis II

Daughter cells of mitosis

Daughter cells of meiosis

© Infobase Learning

FIGURE 7.3 Mitosis is a process that results in the formation of two new cells, each having the same number of chromosomes as the parent cells. Meiosis consists of two divisions (meiosis I and meiosis II) and results in four daughter cells each containing half of the number of chromosomes found in the parent cell.

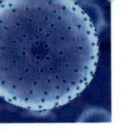

chromatids appear to have a waist at the point where they are most closely attached to each other.

2. *Metaphase:* Cytoskeletal protein cables attach to sister chromatids from two sides of the cell. The protein cables form a structure called the spindle apparatus. Spindle proteins pull chromatids to the cell's equator, which is called the metaphase plate.

3. *Anaphase:* Spindle proteins separate sister chromatids, pulling each member of a pair to opposite sides of the cell, so that each side has a copy of the cell's DNA. After separation, sister chromatids are called chromosomes.

4. *Telophase:* Chromosomes reach the two poles of the cell, and nuclear envelopes form, enclosing chromosomes. Mitosis has ended.

During cytokinesis in an animal cell, a ring of proteins surrounds the cell and contracts, pulling like a drawstring. The contracting ring tightens the plasma membrane and creates a furrow around the cell. When the plasma membrane meets itself in the middle of the cell, it fuses and splits the old cell into two cells, each with its own nucleus and supply of cytoplasmic organelles. Cytokinesis proceeds differently in plant cells, which have a sturdy cell wall that surrounds the plasma membrane. In plant cells, a dividing cell assembles a new cell wall within the cell and between the two sets of chromosomes. As the new wall expands in two directions, it meets the parent cell's plasma membrane and cell wall, splitting the parent cell into two daughter cells.

CELL DIVISION THAT GENERATES GENETIC DIVERSITY

Reproduction by binary fission typically produces a population of identical cells. This can create a risky situation. A toxin that kills one bacterium will kill an entire colony of identical bacteria. The value of genetic diversity is highlighted by the fact that bacteria have evolved mechanisms for shuffling genes to produce bacteria with traits from different cells. One method for exchanging genetic material is conjugation, in which two bacteria attach to each other, whereupon their chromosomes exchange genes. Bacteria can scavenge new genes by absorbing bacterial DNA fragments from the environment and integrating the DNA fragments into their chromosomes. Bacteria also obtain new traits from viruses that infect bacteria

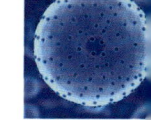

RESISTANCE IS NOT FUTILE

For her doctoral research at the Norwegian School of Veterinary Science, Anne-Mette R. Grønvold investigated changes in the intestinal bacteria of healthy dogs that had been treated with antibiotics. In 2010, she reported that a few days of drug treatment resulted in a change in the types of bacteria that populated the intestines and caused *E. coli* bacteria to develop a resistance to different types of antibiotics. Grønvold also studied the effect of treating horses and calves with the antibiotic penicillin. Again, a few days of drug treatment caused intestinal *E. coli* to become resistant to several types of drugs.

Various mutations can enable a bacterial cell to survive drug treatment. For example, the bacterial cell molecule that is the drug target may be altered, so that it does not bind to the drug. A bacterium can decrease the amount of a drug that enters the cell, or efficiently pump the drug molecules from the cell. A bacterial cell may acquire enzymes that inactivate or destroy the drug. Drug resistance is a result of gene mutation. Sometimes, a bacterium acquires a mutation in its DNA that confers drug resistance. DNA molecules called plasmids that carry drug resistance genes provide another way for a bacterial cell to survive antibiotic treatment. Bacteria can obtain a plasmid from other bacteria. Bacterial cells also acquire drug resistance genes as they scavenge DNA fragments from dead bacteria. A DNA fragment that includes a drug resistance gene can become inserted into the scavenger's chromosomal DNA or into a plasmid that the bacterium carries.

When a bacterial population is exposed to an antibiotic, many bacteria may die, but the resistant cells survive and reproduce. Over time, the selective pressure of drug treatment results in a population of bacteria that are not killed by the drug. Scientists suggest that antibiotic treatment of humans and animals can be viewed as a long term experiment to test Charles Darwin's survival of the fittest hypothesis as it applies to bacteria. The worldwide increase in drug-resistant bacteria supports Darwin's idea.

and from plasmids, which are small, circular DNA molecules that replicate themselves in bacterial cells and can transfer from one bacterial cell to another.

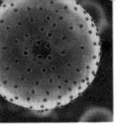

Eukaryotes, from single-celled protists to multicellular plants and animals, multiply by sexual reproduction, using the process of meiosis, which is another type of cell division. In many eukaryotes, cells divide by meiosis to form egg cells and sperm cells. Meiosis proceeds in two stages to create four daughter cells. The first stage of meiosis is similar to mitosis. The nuclear membrane breaks down, exposing the duplicated genome in the form of sister chromatids. Web-like proteins of the spindle apparatus attach to sister chromatids and pull them toward the middle of the cell. An important difference between mitosis and meiosis concerns the way that the spindle proteins divide genetic material into the two sides of the cell. In mitosis, each side of the cell receives a copy of the original genetic material. In the first stage of meiosis, spindle proteins do not divide chromatids into two identical sets.

As an example, suppose that a somatic cell contains only one type of chromosome, called chromosome 1. The cell would have two versions of the chromosome—one from the mother and one from the father. Call them chromosome 1m and chromosome 1f, respectively. The cell gets ready to divide, and its DNA duplicates. Now, the cell contains two copies of chromosome 1m and two copies of chromosome 1f. In mitosis, each side of the cell gets one copy of chromosome 1m *and* one copy of chromosome 1f. When the cell divides, the two daughter cells contain identical DNA. That is, each daughter cell has one copy of chromosome 1m and one copy of chromosome 1f. In the first stage of meiosis, each side of the cell gets two copies of chromosome 1m *or* two copies of chromosome 1f. When the cell divides, the daughter cells do not contain identical DNA.

What does this say about the first stage of meiosis in a human cell? Humans store their genetic information in the DNA of 24 different chromosomes: 22 chromosomes identified by number from largest to smallest, an X chromosome, and a Y chromosome. Egg cells and sperm cells have only 23 chromosomes each. At conception, an egg cell (which has chromosomes 1 to 22 and an X chromosome) and a sperm cell (which has chromosomes 1 to 22 and either an X chromosome or Y chromosome) fuse to form a cell that has the full set of 46 chromosomes—23 from each parent. A typical female has inherited one X chromosome from each parent, whereas a typical male has inherited an X chromosome from his mother and a Y chromosome from his father. Human somatic cells have a nucleus that contains 23 pairs of chromosomes, or 46 individual chromosomes.

Consider a female human cell. DNA duplicates to create the following chromosomes:

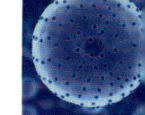

- two copies of chromosomes 1 to 22 inherited from the mother,
- two copies of chromosomes 1 to 22 inherited from the father,
- two copies of the X chromosome inherited from the mother, and
- two copies of the X chromosome inherited from the father.

After the first stage of meiosis, each daughter cell will have two copies of chromosomes 1 to 22 and two copies of an X chromosome. Each cell contains some chromosomes inherited from the mother and some chromosomes inherited from the father.

To complete the picture, consider a male human cell dividing in the first stage of meiosis. In this case, DNA duplicates to create the following chromosomes:

- two copies of chromosomes 1 to 22 inherited from the mother,
- two copies of chromosomes 1 to 22 inherited from the father,
- two copies of the X chromosome inherited from the mother, and
- two copies of the Y chromosome inherited from the father.

After the cell divides, each daughter cell will have two copies of chromosomes 1 to 22—some of them inherited from the father and some of them inherited from the mother. One daughter cell will contain two copies of the X chromosome, and the other will contain two copies of the Y chromosome.

During the second stage of meiosis, each daughter cell divides its set of chromosomes equally between two cells, so that each cell contains the same DNA. Each of the four daughter cells contains half the number of chromosomes of the original parent cell. In males, the four cells will develop into sperm cells, while the cells develop into egg cells in females.

Scientists have divided the processes of meiosis into two stages; each stage has four steps:

Meiosis I

1. *Prophase I:* The nuclear membrane breaks down and the spindle apparatus forms. Structures of four chromatids can be seen under the microscope. A set of four chromatids consists of two sets of sister chromatids—one set inherited from the mother and one set from the father. For example, one group of four chromatids consists of two copies of maternal chromosome 1

and two copies of paternal chromosome 1. Arms from maternal and paternal chromatids overlap and fuse, forming an X-shaped structure called a chiasma (plural: chiasmata).

2. *Metaphase I:* Spindle proteins pull sets of four chromatids—bound together by chiasmata—to the equator of the cell. At the end of metaphase I, chiasmata disband, resulting in an exchange of genes between maternal and paternal chromatids. Chiasmata formation is another process that creates genetic diversity.

3. *Anaphase I:* Spindle proteins contract and separate maternal and paternal sister chromatids toward the two halves of the cell. For example, the sister chromatid pair of chromosome 1 inherited from the mother moves to one end of the cell, and

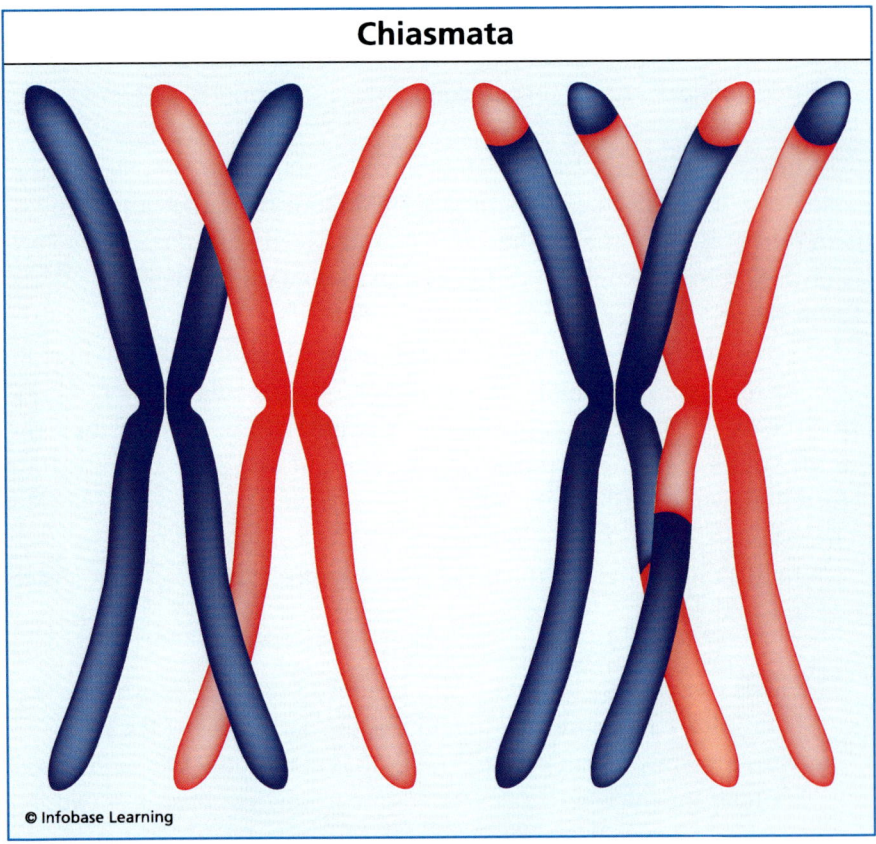

Chiasmata

© Infobase Learning

FIGURE 7.4 Before crossing over, sister chromatid pairs line up. After crossing, chromosomes have exchanged genetic material.

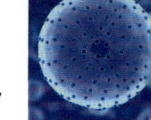

the sister chromatid pair of chromosome 1 inherited from the father moves to the other end of the cell.

4. ***Telophase I and cytokinesis***: Nuclear envelopes enclose chromatids. A groove forms in the middle of the cell and splits the cell into two daughter cells. Each daughter cell contains a set of chromosomes in the form of sister chromatids.

Meiosis II

1. ***Prophase II***: The spindle apparatus forms and begins to pull sister chromatids toward the middle of the cell.
2. ***Metaphase II***: Chromatids align at the equator of the cell.
3. ***Anaphase II***: Spindle proteins separate sister chromatids, so that each side of the cell receives one set of chromosomes.
4. ***Telophase II***: Nuclear envelopes surround chromosomes. A groove forms in the middle of the cell, and cytokinesis creates daughter cells.

Meiosis produces daughter cells that contain half the number of chromosomes found in somatic cells. A reduction in chromosome number is crucial. If meiosis did not halve the number of chromosomes, then the number of chromosomes would double with each generation.

APOPTOSIS: PROGRAMMED CELL DEATH

Necrosis is the death of cells due to an injury, such as a wound, or from an infection by viruses or bacteria. The plasma membrane of a cell undergoing necrosis no longer functions as a selectively permeable barrier. Water rushes into the cell, and the cell bursts, releasing hazardous enzymes and chemicals that damage neighboring cells.

Apoptosis is another type of cell death, a death that is planned and proceeds in an orderly manner. This type of cell death is also called programmed cell death. A cell undergoing apoptosis shrinks, its DNA fragments, its cytoskeleton collapses, and pieces of the cell break away until the cell disintegrates. As the cell falls apart, the plasma membrane fractures into vesicles that enclose cell contents. Phagocytic cells engulf and destroy these biohazard bags. Unlike necrosis, apoptosis does not damage adjacent cells and leaves a neater mess for cleanup.

In a multicellular animal, apoptosis plays vital roles. The combination of apoptosis and mitosis ensures the renewal of tissues. For instance, worn

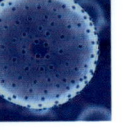

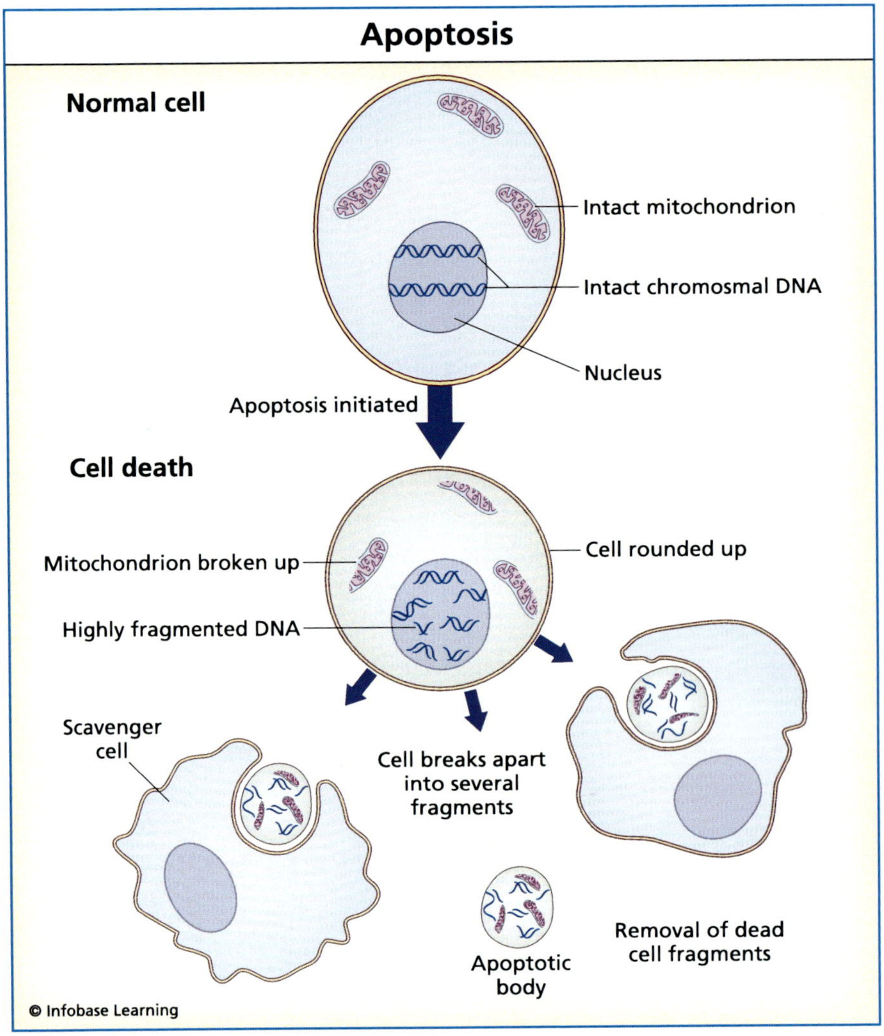

Apoptosis

Normal cell

Intact mitochondrion

Intact chromosmal DNA

Nucleus

Apoptosis initiated

Cell death

Mitochondrion broken up

Cell rounded up

Highly fragmented DNA

Scavenger cell

Cell breaks apart into several fragments

Apoptotic body

Removal of dead cell fragments

© Infobase Learning

FIGURE 7.5 In apoptosis, a cell shrinks as its DNA fragments, the cytoskeleton collapses, and pieces of the cell break away until the cell disintegrates.

cells of the intestine are regularly replaced with new, healthy cells. In an adult human, billions of intestinal cells are replaced every hour.

Animal cells regulate apoptosis with a family of proteases, which are enzymes that cleave proteins. Normally, these proteases are inactive. If an event activates the enzymes, they cleave other proteases to activate them, setting off a cascade of proteolytic activity. Some activated proteases attack proteins crucial for the cell, such as nuclear lamins, which are

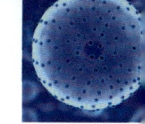

proteins that support the nuclear envelope. Other proteases activate enzymes that cleave DNA.

Signals to begin apoptosis can arrive from other cells or can originate from within a cell. A signal to initiate apoptosis from outside a cell can stimulate death receptors in the plasma membrane. Certain white blood cells, for example, secrete a signaling molecule called Fas ligand that binds with death receptors. A damaged cell may initiate its own apoptosis by causing mitochondria to release a protein that activates proteases. For instance, the cell cycle checkpoint mechanism can transmit a signal to initiate apoptosis if newly synthesized DNA is damaged beyond repair.

EXTRACELLULAR SIGNALS REVISITED

Animal cells receive signals—typically, in the form of proteins—that tell the cells to grow, divide, or die. Examples of these signaling molecules include growth factors, mitogens, and survival factors.

- Growth factors stimulate a cell to increase its size and mass. Extracellular growth factors bind with plasma membrane receptors and stimulate the cell to increase the rate of synthesis of proteins and other macromolecules, while decreasing the degradation of macromolecules.
- Mitogens stimulate cell division, often by releasing a block that prevents a cell from progressing from the G1 phase of the cell cycle into the S phase. At the site of a wound, for example, blood platelets secrete platelet-derived growth factor, which binds with receptors of nearby cells, stimulating the cells to divide and aid in wound healing.
- Survival factors suppress the intracellular suicide program of apoptosis. These signaling molecules play an important role in the development of the nervous system of mammals. In the mammalian embryo, neurons branch out from the central nervous system toward various organs and tissues. An excess number of neurons search for their target cells. The neuron that first reaches a target cell is rewarded by survival factors secreted by the target cell. Neurons that fail to find their target cells are not exposed to survival factors and undergo apoptosis. This system eliminates up to half of developing neurons.

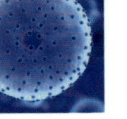

SYNTHIA

In May 2010, researchers at the J. Craig Venter Institute (in La Jolla, California) announced that they had produced the first self-replicating, synthetic bacterial cell. They based the genome of the new cell, dubbed Synthia by some, upon the genome of the bacterium, *Mycoplasma mycoides*. The scientists modified the nucleotide sequences of the genome by adding a few extras, including the encoded names of the researchers, several quotes, and a Web site address. The team assembled the synthetic 1.08 million base pair chromosome of the artificial cell from 1,078 DNA fragments that were 1,080 base pairs long. After constructing Synthia's genome, the researchers inserted the DNA into *Mycoplasma capricolum* bacterial cells. As the bacteria divided, the original bacterial genome was destroyed or lost, leaving the synthetic genome in charge of the cells.

Daniel Gibson, a team leader of the Venter Institute project, explained their accomplishment in a company press release. "To produce a synthetic cell, our group had to learn how to sequence, synthesize, and transplant genomes," he said. "Many hurdles had to be overcome, but we are now able to combine all of these steps to produce synthetic cells in the laboratory." Looking ahead, Gibson said that, "We can now begin working on our ultimate objective of synthesizing a minimal cell containing only the genes necessary to sustain life in its simplest form. This will help us better understand how cells work."

Biotechnology companies are applying the new field of synthetic biology to construct cells that will secrete vaccines, biofuels, pharmaceuticals, and other products. The announcement of a synthetic cell did not please everybody. The environmental advocacy group Friends of the Earth asked the U.S. Environmental Protection Agency and the U.S. Food and Drug Administration to fully regulate any synthetic biology experiments and products. The Canadian ETC Group, which studies social impacts of new technologies, characterized the accomplishment as a "Pandora's Box moment" and urged a global moratorium on synthetic biology.

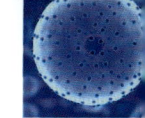

Other extracellular signaling molecules suppress the effects of growth factors, mitogens, and survival factors to limit the size of tissues.

BALANCING CELL DIVISION AND CELL DEATH

Prokaryotes reproduce themselves by binary fission, whereas eukaryotic cells multiply using the process of mitosis. Eukaryotic cells divide by meiosis to create egg cells and sperm cells. An important difference between mitosis and meiosis is that mitosis has one round of DNA replication followed by one cell division, and meiosis has a single round of DNA replication followed by two rounds of cell divisions.

Single-cell organisms usually divide and grow as fast as the supply of nutrients allows. Uncontrolled cell division is incompatible with the requirements of multicellular organisms, in which a cell lives as long as the body needs it. Mitosis is balanced with apoptosis in multicellular animals. On the one hand, too little apoptosis can cause an adult tissue to grow to an inappropriate size. On the other hand, too much apoptosis can produce disease. Nervous system disorders, such as Parkinson's disease and Alzheimer's disease, may result in part from excess apoptosis. Too much mitosis also creates a disease: cancer.

8

Cancer: Unrestrained Cell Division

Transformation is the process that changes a normal cell into a **cancer cell**. Usually, the immune system recognizes and attacks a transformed cell. If an abnormal cell escapes the immune system, it can reproduce, forming a collection of cells—a tumor—within normal tissue. A **benign tumor** is a mass of abnormal cells that remains confined at the location of origin and can often be totally eliminated by surgery. However, if tumor cells invade surrounding normal tissues and spread throughout the body, the tumor is said to be a **malignant tumor**. A person diagnosed with cancer has a malignant tumor.

Any of the many different types of cells in a body can transform into a cancer cell. More than 100 distinct types of cancer have been identified. Cancer cells can be grouped into blood cell cancers and solid tumors. The most common blood cell cancers are leukemias, which arise from cells that form blood cells, and lymphomas, which develop from immune system cells. Scientists classify solid tumors according to the type of cell from which the cancer was derived. In humans, the most common solid tumors are carcinomas, which are cancers derived from epithelial cells. Carcinomas include tumors of the skin, colon, cervix, lung, breast, and other tissues. Sarcomas are cancers of connective and structural tissues, such as bone, cartilage, and muscle. This type of solid tumor is relatively rare in humans.

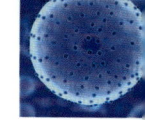

FUNCTIONAL DIFFERENCES BETWEEN A NORMAL CELL AND A CANCER CELL

Regardless of their origin, malignant cancer cells share similar characteristics: unregulated cell division and the ability to invade other tissues. A number of changes in cell activity create these capabilities. The following characteristics enable cancer cells to proliferate indefinitely:

- *Lack of contact inhibition*: Normal cells typically cease cell division when they contact other cells. Cancer cells lack this behavior, which is called contact inhibition.
- *Loss of gap junctions*: In normal cells, gap junctions allow the transfer of molecules that can control cell growth of neighboring cells. Most cancer cells do not have gap junctions.
- *Virtual immortality*: A typical normal cell and its progeny die after a certain number of cell divisions, whereas cancer cells and their progeny can divide indefinitely. Unlike normal cells, cancer cells can resist apoptosis—programmed cell death—and have extended life spans.
- *Independence from external growth-promoting signals*: In the body, cells secrete growth factor proteins that travel through the bloodstream (endocrine stimulation) or diffuse to nearby cells (paracrine stimulation). External growth factors limit the proliferation of normal cells. Cancer cells have reduced requirements for growth factors produced by normal cells. Cancer cells exhibit autocrine growth stimulation, which means that cancer cells synthesize their own growth factors, secrete them, and then respond to the growth factors by producing more cancer cells. Cancer cells do not respond to signal molecules sent by neighboring cells in a tissue. They live independently of their neighbors, cut off from the needs of the tissue and the body.

Metastasis is the ability of a cancer cell to invade other tissues, and it requires a number of changes in cellular function as well. A cancer cell secretes proteases into the extracellular matrix that surrounds the cell. The enzymes degrade the matrix. Changes in the plasma membrane and cytoskeleton of a cancer cell enable the cell to move through the newly created holes in the matrix. Cancer cells also stimulate **angiogenesis**, which is the formation of blood vessels. Under normal conditions,

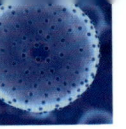

angiogenesis occurs during wound healing and in fetal development. Cancer cells secrete certain growth factor proteins that provoke blood vessel cells to migrate to the cancer cells and to reproduce, forming new blood vessels for the tumor. The new blood vessels not only supply oxygen and nutrients for the cancer cells, but also provide a route for the cells to migrate and colonize new tissues. As they travel throughout the body, many cancer cells evade detection by immune system cells due to changes in their plasma membrane proteins.

GENETIC DIFFERENCES BETWEEN A NORMAL CELL AND A CANCER CELL

A tumor develops from a cell with a mutation that enables a cell to divide independent of the needs of the surrounding tissue cells. As abnormal cells produce a mass, new mutations confer the additional abilities of a cancer cell: prolonged survival, uncontrolled proliferation, and the ability to invade

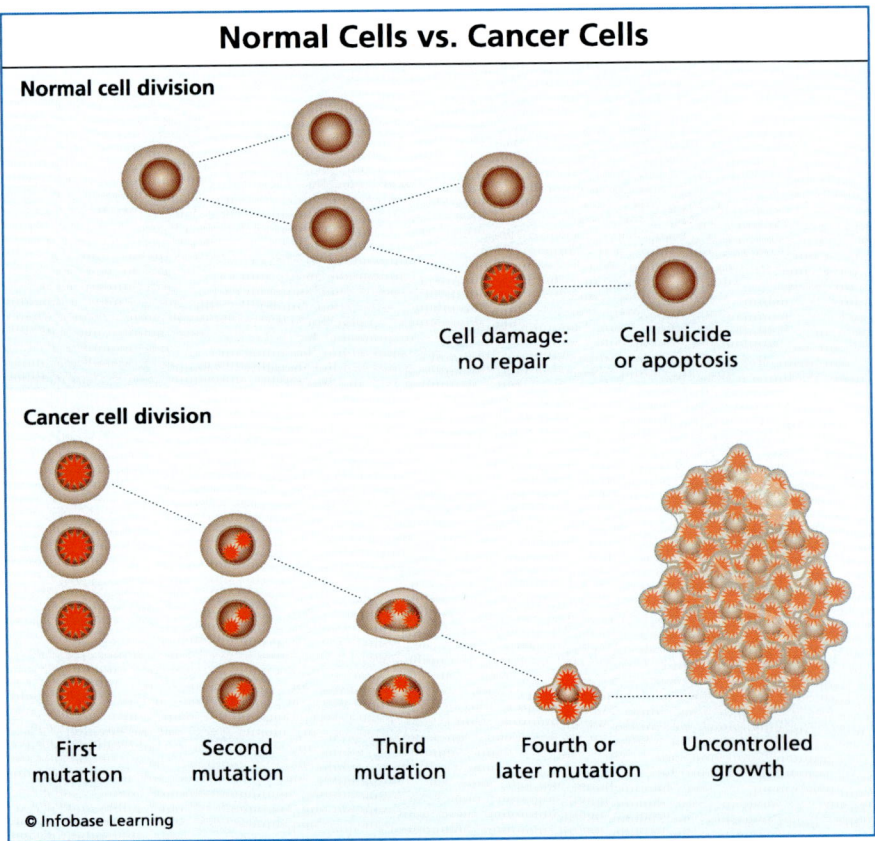

FIGURE 8.1 Cancer cells grow and divide at a fast pace and are considered malignant if the cells invade normal tissue.

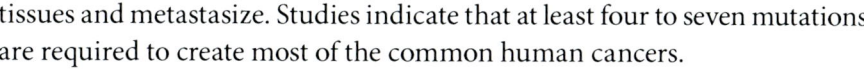

tissues and metastasize. Studies indicate that at least four to seven mutations are required to create most of the common human cancers.

The Nature of Gene Mutations

Before plunging into the topic of gene mutation, a brief review may be helpful. Recall that a gene is a DNA nucleotide sequence that provides data for a cell to synthesize a protein. Messenger RNA molecules carry this information from a DNA molecule to the protein synthesis machinery, where the genetic code enables the translation of an RNA nucleotide sequence into instructions for assembling amino acids into proteins.

The genetic code is based on codons, which are triplets of the four nucleotide bases found in RNA: adenine, guanine, cytosine, and uracil. A coding system based upon a series of three of four possible bases yields 64

THE ROLE OF THE MAHJONG GENE IN A DEADLY CELL COMPETITION

Tumor cells proliferate and invade normal tissue. How tumor cells make room in normal tissue for their continuing invasion has been a mystery. Scientists report that fruit fly tumor cells induce normal, adjacent cells to undergo apoptosis. Then, the tumor cells, winners of the cell competition, spread into the space vacated by the normal cells.

Under certain conditions, normal cells out-compete abnormal cells. Researchers at Florida State University and University College London identified a key gene in cell competition, which they named the Mahjong gene after the Chinese tile game that requires strategy, skill, and chance. Mahjong, the protein encoded by the Mahjong gene, interacts with a tumor suppressor protein called Lgl, which normally controls cell division. Mutations in the Lgl gene promote the transformation of a normal cell into a cancer cell.

In experiments on fruit fly cells, the cells that lacked a Mahjong gene died when they were adjacent to normal cells, but survived if surrounded by other cells that lacked the Mahjong gene. Cells that lacked Mahjong lost the competition for survival with normal cells. Experiments with kidney cells revealed that Mahjong regulates cellular competitiveness in mammalian cells as well. The studies provided the first evidence of genes being involved in mammalian cell competition. Further research into cell competition may suggest new cancer therapies that aid normal cells in the struggle against tumor cells.

(4 x 4 x 4) possible combinations. Cells use all 64 codons, and yet, there are only 20 common amino acids. Some amino acids are encoded by two or more codons. For example, the amino acid leucine is encoded by the codons UUA, UUG, CUU, CUA, CUC, and CUG. Some codons serve as start and stop signals for protein synthesis. The codon AUG encodes the amino acid methionine. AUG also signals the place in the nucleotide sequence of messenger RNA where the code for a protein begins. By marking the point where the protein encoding sequence starts, the AUG codon determines how the sequence should be grouped into triplets. The AUG codon creates the reading frame for the nucleotide sequence. As an example, consider the following the nucleotide sequence:

... GCA AGG CCG AUG GGG CGA AUU GCC UGC CCG UGA ...

The AUG codon creates a reading frame and signals that the nucleotide sequence should be read as:

AUG GGG CGA AUU GCC UGC CCG UGA.

The genetic code signals the end of protein synthesis with a stop codon. In the example above, a UGA codon signals the end of a protein-encoding nucleotide sequence. UAA and UAG are also stop codons. Sometimes, the stop codons are referred to as "nonsense codons."

The addition or deletion of a single nucleotide can create a frameshift mutation. This type of mutation changes the reading frame of a nucleotide sequence that encodes protein. By shifting the way that messenger RNA bases are grouped into three, the mutation alters the series of amino acids. Consider the following short piece of messenger RNA that encodes the amino acid sequence, leucine-valine-alanine-glutamine:

CUU GUU GCU CAA.

Suppose that a mutation in DNA results in the loss of the first uracil in the messenger RNA. The reading frame shifts to:

CUG UUG CUC AA.

The new sequence encodes leucine-leucine-leucine with two adenine bases left over.

Now, suppose that a mutation caused the addition of a base in the original nucleotide sequence. For example, a guanine is added next to the first guanine in the sequence. The mutated sequence would be grouped as:

CUU GGU UGC UCA A.

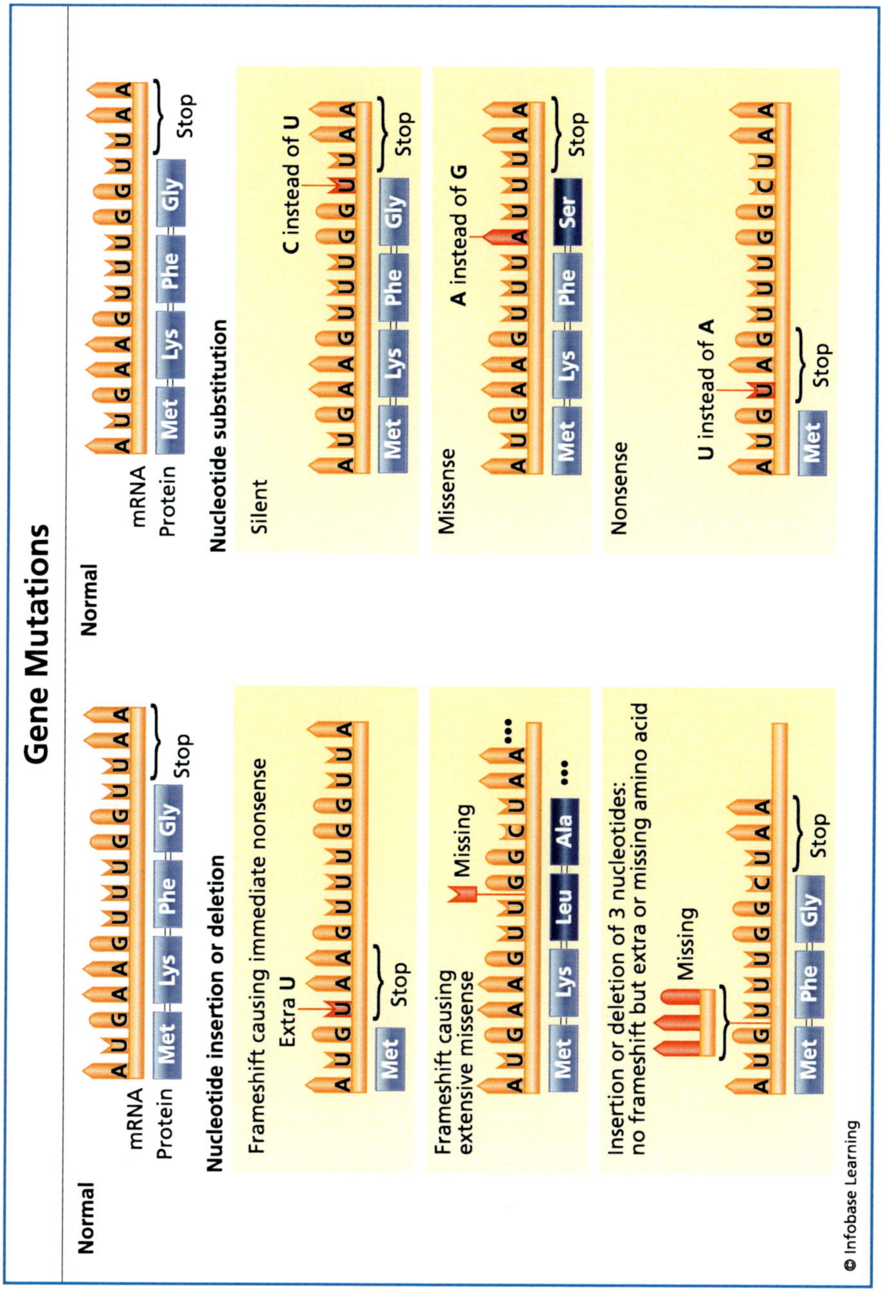

Gene Mutations

Normal

mRNA A U G A A G U U U G G U U A A } Stop

Protein Met — Lys — Phe — Gly

Nucleotide substitution

Silent

A U G A A G U U U G G U U A A } Stop
                    ↑ C instead of U
Met — Lys — Phe — Gly

Missense

A U G A A G U U U A U U U A A } Stop
                    ↑ A instead of G
Met — Lys — Phe — Ser

Nonsense

A U G U A G U U U G G C U A A } Stop
        ↑ U instead of A
Met

Nucleotide insertion or deletion

Frameshift causing immediate nonsense

A U G U A A G U U U G G U U A } Stop
        ↑ Extra U
Met

Frameshift causing extensive missense

A U G A A G U U G G C U A A •••
                    ↑ Missing
Met — Lys — Leu — Ala •••

Insertion or deletion of 3 nucleotides: no frameshift but extra or missing amino acid

A U G U U U G G C U A A } Stop
     ↑↑↑ Missing
Met — Phe — Gly

© Infobase Learning

FIGURE 8.2 Gene mutations can take various forms. For example, a frameshift mutation changes the amino acid sequence of a protein as nucleotides are added or deleted. A silent mutation, on the other hand, will not result in a change to the amino acid sequence.

This sequence encodes leucine-glycine-cysteine-serine with an adenine left over.

Single nucleotide mutations not only cause the loss or gain of a nucleotide: A mutation can also replace one nucleotide with another. A nucleotide replacement can affect a codon in one of three ways:

- In a silent mutation, the mutated codon still encodes the same amino acid. For instance, the mutation of CAA to CAG does not affect the amino acid added to a protein; both codons encode the amino acid glutamine.
- In a nonsense mutation, a nucleotide replacement creates a stop codon from a codon that encoded an amino acid. The mutation of CAA to UAA is a nonsense mutation. Since protein synthesis stops early, a nonsense mutation results in a shortened protein. The short protein may not function or may function poorly.
- In a missense mutation, the mutation alters a codon for one amino acid to a codon for a different amino acid. For example, the mutation of CAA to GAA would replace glutamine with glutamic acid. A missense mutation can have little effect on protein function or it can greatly alter it.

A gene mutation can take many forms. Some mutations are caused by a single base change, while others are caused by a large alteration in a gene, the loss of a gene, the gain of a new gene, or duplication of genes. Sometimes, an alteration of a chromosome is so huge that the change can be seen under a microscope.

Cancer Cells Have Altered Gene Expression

The patterns of gene expression differ in normal cells and in cancer cells. Gene mutation does not account for all of these changes. As discussed in a previous chapter, eukaryotic cells use various methods for controlling gene transcription. One type of control mechanism is DNA methylation, in which an enzyme attaches a methyl group to a cytosine base in a DNA molecule. Attachment of methyl groups can hinder the synthesis of mRNA from DNA and stops the production of the protein encoded by the mRNA. Alterations in the chromosome protein histone also affect gene transcription. Attachment of acetyl groups to histones decreases the positive charge of the proteins and reduces the attraction between histones and negatively charged DNA. Consequently, acetyl group attachment results in the release of histone proteins from DNA, which allows transcription factors to access genes. Active areas of chro-

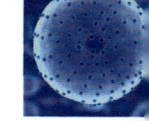

matin can have unmethylated DNA and large amounts of histones with acetyl groups.

Compared with normal cells, cancer cells contain an abundance of unmethylated DNA. Cancer-cell DNA also contains regions of unusually methylated DNA. Hypermethylation at the start sites of transcription results in the silencing of gene transcription. In human cancers, hypermethylated genes include genes that normally suppress the transformation of normal cells into tumor cells. Cancer-cell chromosomes also have modified histones, which result in altered patterns of gene transcription.

Another mechanism that cells use to control protein synthesis is RNA interference, in which micro-RNAs destroy mRNA molecules or block translation. This gene expression control system is also altered in cancer cells. Studies have shown that genes for two micro-RNAs were deleted or expressed at abnormally low levels in a type of leukemia cell. The micro-RNAs normally decrease the synthesis of a protein called BCL2, which prevents a cell from undergoing apoptosis. If the micro-RNAs cannot perform their job, a cell will synthesize too much BCL2 protein. As a result, BCL2 proteins block apoptosis and help a cell to become immortal.

MANY FACTORS LEAD TO GENE MUTATION

A mutation can be an inherited mutation or a sporadic mutation. An inherited mutation exists at the time of conception and is contributed by DNA in the egg cell, the sperm cell, or both cells. Inherited mutations that cause cancer are rare in humans. Sporadic mutations are common. They occur during the life of a cell due to errors in DNA synthesis or DNA repair, or result from external influences from the environment. Regardless of whether mutations are inherited or sporadic, the mutations must alter certain genes to transform normal cells into cancer cells. Scientists have found three groups of genes, which, when mutated, promote the development of cancer: proto-oncogenes, tumor suppressor genes, and DNA repair genes.

Normal cells have multiple control systems that promote cell division, protect against unchecked cell proliferation, and correct gene mutations. Proto-oncogenes promote normal cell division. In contrast, a mutated proto-oncogene, called an **oncogene**, can stimulate unchecked cell proliferation, or block apoptosis. Some oncogenes interfere with the development of normal, specialized activities. For instance, an oncogene may prevent the development of a liver cell, so that the cell retains general functions. **Tumor suppressor genes** protect against uncontrolled cell growth and can promote apoptosis. For example, the tumor suppressor p53 protein stimulates apoptosis if the cell has suffered significant DNA

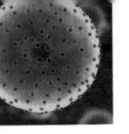

damage. The p53 protein has been altered in most human cancer cells. If mutations convert proto-oncogenes to oncogenes and render tumor suppressor genes nonfunctional, the cell cannot defend itself against uncontrolled proliferation, and the mutations contribute to the transformation of normal cells into cancer cells.

Cells have DNA repair systems to fix mutated DNA. If they miss a mutation, then a safeguard system prevents cells from dividing to allow DNA repair systems another opportunity to fix the damage. If DNA damage is severe, then a safeguard system will initiate apoptosis. Sometimes a mutation disables one of the safeguard systems that repair mutated DNA. With the safeguard system out of order, the mutation rate escalates and the cell accumulates multiple mutations. Defects in the DNA repair system are common in cancer cells, which have continuously mutating genes.

Gene mutations occur in many ways. Sometimes, an error occurs when DNA is synthesized for cell division, and the uncorrected error is passed on to daughter cells. Various chemicals in food and in polluted air also induce mutations. A chemical that promotes the development of cancer is a carcinogen. Many industrial chemicals, such as benzene and arsenic, are carcinogens. Cigarette smoke includes at least 40 types of carcinogens. Radiation also inflicts mutations. A common form of mutation-causing radiation is ultraviolet light from the sun, which promotes the development of skin cancer.

In their 1983 dictionary, *Aristotle to Zoos*, scientists Peter and Jane Medawar define a virus as a piece of bad news wrapped in protein. The bad news part of the virus is the viral genome in the form of a DNA or RNA molecule. The nucleic acid molecule contains instructions that enable a virus to take over a host cell, and force the cell to produce proteins and nucleic acid molecules for new viruses.

Many types of viruses infect humans and certain types are associated with cancer. Although viruses do not cause cancer by themselves, they can accelerate the transformation process. Some viral genomes encode proteins that interfere with the safeguarding systems of normal cells. For example, the Hepatitis B virus infects liver cells and inactivates the p53 system, a defense against inappropriate cell division. With this safeguard neutralized, cells require fewer mutations to transform into cancer cells. Viruses can also carry oncogenes or control a cellular proto-oncogene after infection.

Sometimes, a person is born with a predisposition for a certain type of cancer. Retinoblastoma, for example, is a rare childhood cancer of the retina of the eye, which can begin with inherited deletions in chromosome 13 and the loss of one copy of the RB gene. Studies indicate that the RB protein helps to regulate the passage of cells from the G1 phase of the cell cycle to the

HOT WATER!

Today, few would consider taking a relaxing soak in a tub of radioactive water. During the early 1900s, however, exposure to radioactive water became a popular health fad. For example, promoters of a hot spring resort located in Arkansas discovered the presence of radioactive radium in the water. Rather than isolating the area with dire warnings, they boasted about the contamination. Radioactive substances carry electrical energy deep into the body, they told the public, and this energy would not only stimulate organs at the cellular level, but also hasten the departure of waste from the body. In response, health-conscious people flocked to the hot springs.

In time, companies cashed in on the radioactivity craze by making products that allowed consumers to bring radioactive sources into their homes. Radium water dispensers became very popular during the 1920s. "Drink liquid sunshine," companies advised, "and expose your internal organs to the sun's healing rays." One company, the Revigator Water Jar Company of San Francisco, sold several hundred-thousand Revigator radioactive water dispensers. To promote their product, the company published a red-covered booklet entitled *The Revigator Water Jar for Every Home*, in which they claimed that a "surprising amount of illness is caused from drinking improper water." The company explained that natural water discharged from the earth is radioactive, whereas ordinary drinking water does not possess that property to any significant extent. The Revigator was said to remedy this situation. The jug was lined with a uranium ore, and released radon gas that dissolved in the stored water. In this way, the booklet stated, the "ore continuously revigorates or restores natural vigor to drinking water placed therein." According to the company, Revigator water tasted softer and more palatable. While some customers guzzled water from these uranium-lined jugs, others preferred to get their "liquid sunshine" from elixirs of radioactive radium in distilled water.

By the 1930s, scientists and physicians realized that radiation can alter human cells and promote the development of cancer. Fears about radioactive contamination gradually replaced the general public's fascination with the supposed health effects of radioactive water.

S phase. After a cell enters the S phase, it typically proceeds to mitosis. The RB protein serves as a brake on the progression into the S phase. Before normal cells transform into retinoblastoma cells, both maternal and paternal

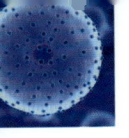

copies of the gene must be lost or mutated. A child who inherits a chromosome with a deleted RB gene is already half way there.

CANCER THERAPIES TARGET UNIQUE TRAITS OF CANCER CELLS

A solid tumor can be surgically removed from a patient. Surgical treatment may also be used in combination with radiation therapy, in which high-energy radiation targeted at a tumor damages DNA and blocks DNA replication. The rationale of radiation therapy is that most normal adult cells do not divide frequently. In contrast, cancer cells divide continuously and they often have defective DNA repair systems. Physicians use other tactics to treat patients who have blood cell cancers and malignant cells that have spread from a solid tumor. Traditionally, physicians have used chemotherapy and radiation therapy. With varying degrees of success, these treatments kill cancer cells while allowing normal cells to survive. Chemotherapy is based upon the administration of toxic chemicals that kill rapidly growing cells. Some drugs interfere with DNA synthesis, while other drugs prevent the completion of mitosis. For example, the anti-cancer drug Paclitaxel (also known as Taxol®) blocks the activities of proteins that make up the mitosis spindle apparatus and freezes cells in the M phase of the cell cycle. Again, the basis of chemotherapy is that most normal adult cells do not divide frequently, unlike cancer cells. However, some normal adult cells do proliferate continuously, such as the cells that line the gastrointestinal tract. Nausea is one common side effect of chemotherapy.

Newer therapies more specifically target cancer cells and may not harm normal cells. These treatments aim at particular characteristics of cancer cells:

- Angiogenesis inhibitors prevent the promotion of blood vessel growth by tumor cells.
- Certain cancer cells require hormones for growth. Hormone therapy aims to deprive cancer cells of the hormones. Some breast cancer cells, for example, require estrogen. The growth of estrogen-dependent tumors can be slowed by preventing estrogen from binding to estrogen receptors in cancer cells.
- Immunotherapies offer another approach. One type of immunotherapy uses antibodies that bind with a protein synthesized by cancer cells but not by normal cells. Chemotherapy drugs or radioactive molecules are attached to the antibodies so that the antibodies can deliver toxins directly to cancer cells. In another

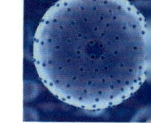

type of immunotherapy, antibodies bind with and inactivate tumor-specific proteins required for survival of the cells. For example, certain breast cancer cells produce large amounts of a certain growth factor receptor protein that stimulates tumor growth. Herceptin® is an antibody that binds with the receptor protein and inhibits its function.

- Gleevec® (also called imatinib) is a new type of cancer therapy that stimulates tumor cells to undergo apoptosis. The drug blocks the activity of an oncogene protein, which promotes uncontrolled growth and reproduction of certain cancer cells.

As scientists learn more about the differences between normal cells and cancer cells, they can design new therapies that specifically target transformed cells.

CANCER CELLS ARE SIMILAR TO SINGLE-CELL ORGANISMS

Cancer is a disease of multicellular organisms. Recall the five categories of cell cooperation, starting at the highest level of organization:

- Organs work together to create organ systems.
- Tissues form organs.
- Specialized cells form tissues.
- Specialized cells work together.
- Cells survive individually.

When a normal cell in a tissue transforms into a cancer cell, it "devolves" through these categories of cellular cooperation. Cut off from communication with other cells, a tumor cell no longer participates in an organ system or as a part of a tissue. A transformed cell loses specialized functions and invests its chemical energy in general activities required for the cell to survive and multiply.

The collapse of bridges and buildings, catastrophic spills of industrial toxic chemicals, and other engineering disasters usually result from a combination of multiple factors, such as design flaws, unusual environmental conditions, failure of materials, and accidents caused by humans. Similarly, the malignant transformation of normal cells into wildly proliferating cancer cells requires a combination of genetic mutations and alterations in gene expression. These changes overcome the many control systems that efficiently safeguard the normal and complex activities of a cell.

Glossary

adaptive immunity (specific immunity) Response of the immune system to antibodies and white blood cells that target foreign (non-self) antigens

adenosine triphosphate (ATP) An activated carrier molecule that shuttles chemical energy within a cell

alternative splicing The process of producing different mature messenger RNA (mRNA) molecules from the same primary RNA transcript by connecting exons in different patterns

amino acid The chemical building block of a protein

anabolic A series of chemical reactions in which complex molecules are synthesized from simpler chemicals with the consumption of chemical energy

angiogenesis The formation of blood vessels

anticodon A sequence of three nucleotides on transfer RNA (tRNA) that can form base pairs with a codon on messenger RNA (mRNA)

antigen A substance that the immune system recognizes as foreign

apoptosis Programmed cell death

base A molecule that forms part of DNA and RNA

base pair Two bases from two nucleotides, held together by weak bonds, in a double-stranded DNA or RNA molecule

benign tumor A mass of abnormal cells that remains confined at the location of origin

binary fission A type of prokaryotic cell division that typically produces two identical daughter cells

cancer cell A eukaryotic cell that divides without control and can invade other tissues

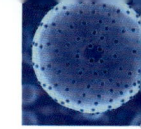

carbohydrate   Compound consisting of carbon, hydrogen, and oxygen atoms

catabolic   A series of chemical reactions in which complex molecules are broken down into simpler chemicals with the generation of chemical energy

cell cycle   Stages in the life of a cell from one cell division to the next cell division

cell-tissue level of organization   A level of cooperation between cells in which similar cells are assembled into tissues

cellular level of organization   A level of cooperation between cells characterized by a division of labor among specialized cells that live together and perform specific functions

chemical energy   Energy stored in chemical bonds that can be released in a chemical reaction

chloroplast   Plant organelle that uses solar energy to produce complex sugar molecules, high-energy chemicals, and oxygen from water and carbon dioxide

chromatin   A mixture of proteins and DNA

chromosome   A structure in a cell that contains DNA

codon   A group of three nucleotides. Most codons code for an amino acid.

complementary   The relationship between nucleotides in two DNA strands or between the nucleotides of a DNA molecule and an RNA molecule. Nucleotide sequences are said to be complementary if the nucleotides of both strands can form base pairs.

covalent bond   A chemical link created when two atoms of different chemicals share electrons

cytoplasm   The organized complex of fluid and organelles outside the nuclear membrane of a cell

cytoskeleton   The network of long proteins that controls cell shape and enables movement

cytosol   The fluid material outside the nucleus and the organelles

deoxyribonucleic acid (DNA)   A nucleic acid molecule that encodes genetic information and contains deoxyribose sugar

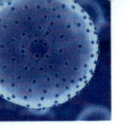

deoxyribose   A five-carbon sugar called ribose that is missing an oxygen atom (deoxy-) on its second carbon and is part of a nucleotide that makes up DNA molecules

disaccharide   A sugar molecule consisting of two monosaccharides joined by a covalent bond

endoplasmic reticulum   A system of membranes in the cytoplasm and a site of protein synthesis

enzyme   A protein that increases the rate of a chemical reaction

eukaryotic cell   A cell that contains a nucleus

exon   Nucleotide sequences that occur in DNA and in pre-mRNA transcripts that encode portions of a protein. Exons of a pre-mRNA transcript are spliced together to form the final nucleotide sequence of mRNA for translation into protein.

extracellular matrix   A mixture of proteins and carbohydrates that surrounds animal cells

gamete   An egg cell or a sperm cell

gap junction   Channels between adjacent animal cells that allow molecules to pass from the cytoplasm of one cell to the cytoplasm of a neighboring cell

gene   A nucleotide sequence that encodes a protein or a functional RNA molecule, such as tRNA

gene expression   Process in which information stored in a DNA molecule is used to make a product, such as a protein

genetic code   The collection of 64 codons that specify 20 amino acids and the signals for stopping protein synthesis

genome   The complete set of an organism's genes

glycoprotein   A protein with attached carbohydrates

Golgi complex   Collection of disk-shaped cytoplasmic organelles that transport protein

histone   Protein associated with DNA in chromatin

hydrophilic   To have an affinity for water

hydrophobic   To lack an affinity for water and have a tendency to repel water

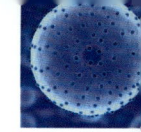

innate immunity (nonspecific immunity)   An inborn ("hard-wired") group of immune defenses against toxins and infectious agents that have been identified as foreign to the body

intron   Nucleotide sequences that occur in DNA and in pre-mRNA transcripts and that do not encode portions of a protein. To form mRNA for translation into protein, introns of a pre-mRNA transcript are cut out and exons are spliced together.

lipid   A varied collection of nonpolar biological molecules, which includes neutral fats, phosopholipids, and steroids

lysosome   A cellular organelle that contains enzymes that break down complex chemicals for reuse by the cell

malignant tumor   A mass composed of abnormal cells that invade surrounding normal tissues and spread throughout the body

meiosis   A type of cell division required to produce egg cells and sperm cells

messenger RNA (mRNA)   An RNA molecule that transmits genetic information from DNA to a cell's protein-making apparatus

metabolism   Life-sustaining chemical processes

metastasis   The ability of a cancer cell to invade other tissues and move to various areas of the body

micro-RNAs   Short RNA molecules that play a role in controlling gene expression

mitochondria   Organelles that function as a cell's power plant

mitosis   A type of eukaryotic cell division that typically produces two identical daughter cells

monosaccharide   "Simple sugars," which in cells have the molecular formula $(CH_2O)_n$, where n has a value of 3 to 7

mutation   A change in the nucleotide sequence of a DNA molecule or a change in the amino acid sequence of a protein

neutral fat   A type of lipid that provides an important source of fuel for animals

nonpolar biological molecules   Molecule that are hydrophobic

nuclear envelope   A membrane that separates the contents of the nucleus from the cytoplasm

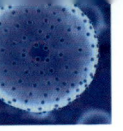

nucleic acid   Any nucleotide chain(s) of DNA or RNA

nucleotide   The monomer of DNA which contains a sugar molecule, a chemical group that contains phosphorus, and a base

nucleus   The organelle that contains most of a cell's DNA

oncogene   A gene encoding a protein that promotes unregulated cell division, and other characteristics of a cancer cell

operator   A component of an operon in a bacterial chromosome to which an active repressor can bind

operon   A segment of DNA in a bacterial chromosome, comprising a promoter, an operator, and structural genes

organ-system level of organization   A level of cooperation between cells in which organs work together to create an organ system

organelle   A membrane-bound structure that performs a function within a cell

peptide   A short chain of linked amino acids

peroxisome   Organelle that purges toxic substances from a cell

phospholipid   A type of lipid that has both hydrophobic and hydrophilic regions and is a vital component of the plasma membrane of animal cells

photosynthesis   Process that uses energy from sunlight to create organic (carbon-containing) molecules from atmospheric carbon dioxide

plasma membrane   A membrane that surrounds a cell

polar molecules   Molecules that are hydrophilic

polymer   A large chemical made by combining smaller chemical units

polysaccharide   A polymer of sugar molecules

primary (protein) structure   Sequence of amino acids in a protein

primary RNA transcript   Product of transcription in eukaryotic cells, which is processed into mRNA

prokaryote   A single-celled organism that lacks a nucleus

promoter   Segment of DNA where RNA polymerase binds to begin synthesis of mRNA

protein   A polymer of amino acids

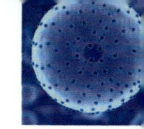

protein synthesis   Production of a protein polymer formed by the addition of amino acids that connect with each other by covalent bonds

protoplasmic level of organization   A type of existence represented by single-celled organisms, in which life-sustaining functions occur within a cell

quaternary structure   Complex of two or more similar or dissimilar protein subunits that form an active molecule

reading frame   Grouping of nucleotides into triplets that may encode a protein

receptor   A complex molecule, such as a protein, that binds with a small molecule in which the binding results in signal transduction. Protein receptors are located in the plasma membrane or within the cell.

ribonucleic acid (RNA)   A nucleic acid molecule that can encode genetic information and contains ribose sugar

ribose   A five-carbon sugar molecule that forms part of a nucleotide that makes up RNA molecules

ribosome   Structure within a cell that contains proteins and RNA and enables the translation of mRNA

RNA interference (RNAi)   A process that decreases or blocks protein synthesis

RNA polymerase   Enzyme that synthesizes mRNA from a DNA strand

secondary structure   Helical or sheet-like structures formed by hydrogen bonding within a protein chain

signal transduction   Process in which the binding of a small molecule to a receptor protein initiates a series of chemical changes that results in an altered cell activity, such as a modification of metabolism

somatic cell   A cell other than an egg cell or a sperm cell

steroid   A type of lipid that has the structure of a complex alcohol, such as cholesterol, vitamin D, and estrogen

structural gene   Gene that encodes a protein

tertiary structure   Three-dimensional shape of a protein formed by the arrangement of secondary structures

tissue-organ level of organization   A level of cooperation between cells in which tissues assemble into organs

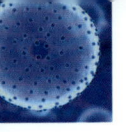

transcription   The process of making an RNA copy of a DNA nucleotide sequence

transfer RNA (tRNA)   An RNA molecule that enables protein synthesis by binding to mRNA and an amino acid

transformation   The process that changes a normal cell into a cancer cell

translation   The process of making a protein using genetic code information in messenger RNA

tumor suppressor gene   A gene encoding a protein that protects against uncontrolled cell proliferation

wobble   The ability of a tRNA molecule to bind with more than one type of base triplet in mRNA that encodes the same amino acid

Bibliography

Abbas, Abul K., Andrew H. Lichtman, and Jordan S. Pober. *Cellular and Molecular Immunology,* 4th ed. New York: W.B. Saunders Company, 2000.

Alberts, Bruce, Dennis Bray, Karen Hopkin, Alexander Johnson, Julian Lewis et al. *Essential Cell Biology,* 3rd ed. New York: Garland Science, 2009.

Altman, Lawrence K. "Team Creates Rat Heart Using Cells of Baby Rats." *The New York Times.* January 14, 2008.

Bennett, P.M. "Plasmid Encoded Antibiotic Resistance: Acquisition and Transfer of Antibiotic Resistance Genes in Bacteria." *British Journal of Pharmacology* 153 (2008): S347–S357.

Campbell, Neil A., Jane B. Reece, Lisa A. Urry, Michael L. Cain, Steven A. Wasserman, Peter V. Minorsky, and Robert B. Jackson. *Biology*, 8th ed. San Francisco: Pearson/Benjamin Cummings, 2008.

Charlesworth, Jac C., Joanne E. Curran, Matthew P. Johnson, Harald H.H. Göring, Thomas D. Dyer et al. "Transcriptomic Epidemiology of Smoking: The Effect of Smoking on Gene Expression in Lymphocytes." *BMC Medical Genomics* 3 (2010). Available online. URL: http://www.biomedcentral.com/1755-8794/3/29. Accessed July 31, 2010.

Chen, Yuang-Tsong. "Glycogen Storage Diseases and Other Inherited Disorders of Carbohydrate Metabolism." In *Harrison's Principles of Internal Medicine,* 16th ed. Dennis L. Kasper, Eugene Braunwald, Anthony S. Fauci, Stephen L. Hauser, Dan L. Longo, and J. Larry Jameson, 2319–2323. New York: The McGraw-Hill Companies, Inc., 2005.

Cooper, Geoffrey M. and Robert E. Hausman. *The Cell: A Molecular Approach,* 5th ed. Washington, D.C.: ASM Press, 2009.

Cree, L.M., D.C. Samuels, and P.F. Chinnery. "The Inheritance of Pathogenic Mitochondrial DNA Mutations." *Biochim Biophys Acta* 1792 (2009): 1097–1102.

Devlin, Thomas M., ed. *Textbook of Biochemistry with Clinical Correlations,* 7th ed. New York: John Wiley & Sons, Inc., 2011.

DiVernieri, Rosanne. "Stinks and Bangs: The Heyday of the Chemistry Set." *Endeavour* 32 (2008): 107–110.

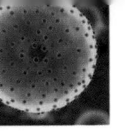

"Divide and Conquer: Genes Decide Who Wins in the Body's Battle Against Cancer." *ScienceDaily* (July 14, 2010). Available online. URL: http://www.sciencedaily.com/releases/2010/07/100713171605.htm Accessed July 18, 2010.

Eliasson, Lena, Fernando Abdulkader, Matthias Braun, Juris Galvanovskis, Michael B. Hoppa, and Patrik Rorsman. "Novel Aspects of the Molecular Mechanisms Controlling Insulin Secretion." *Journal of Physiology* 586 (2008): 3313–3324.

Evans-Molina, Carmella, James C. Garmey, Robert Ketchum, Kenneth L. Brayman, Shaoping Deng, and Raghavendra G. Mirmira. "Glucose Regulation of Insulin Gene Transcription and Pre-mRNA Processing in Human Islets." *Diabetes* 56 (2007): 827–835.

"Everlasting Antibiotics." *Science Illustrated* (May/June 2010): 58–59.

Fierer, Noah, Christian L. Lauber, Nick Zhou, Daniel McDonald, Elizabeth K. Costello, and Rob Knight. "Forensic Identification Using Skin Bacterial Communities." *Proceedings of the National Academy of Science USA* 107 (2010): 6477–6481.

"First Self-Replicating Synthetic Bacterial Cell," J. Craig Venter Institute Web site (May 20, 2010). Available online. URL: http://www.jcvi.org/cms/press/press-releases/full-text/article/first-self-replicating-synthetic-bacterial-cell-constructed-by-j-craig-venter-institute-researcher. Accessed November 27, 2010.

"Grains of Truth." *Science Illustrated* (July/August 2010): 36–45.

Griffiths, Anthony J.F., Susan R. Wessler, Richard C. Lewontin, and Sean B. Carroll. *Introduction to Genetic Analysis,* 9th ed. New York: W.H. Freeman and Company, 2008.

Hartwell, Leland H., Leroy Hood, Michael L. Goldberg, Ann E. Reynolds, Lee M. Silver, and Ruth C. Veres. *Genetics: From Genes to Genomes,* 3rd ed. New York: McGraw-Hill, 2008.

Hickman, Cleveland P. Jr., Larry S. Roberts, Allan Larson, Helen L'Anson, and David J. Eisenhour. *Integrated Principles of Zoology,* 13th ed. New York: McGraw-Hill, 2006.

Hotz, Robert Lee. "Scientists Create Synthetic Organism." *The Wall Street Journal*, May 21, 2010.

"Impact of Antibiotic Treatments on Bacteria in the Intestines of Animals." *ScienceDaily* (April 13, 2010). Available online. URL: http://www.sciencedaily.com/releases/2010/04/100413081238.htm. Accessed July 20, 2010.

Karp, Gerald. *Cell and Molecular Biology*, 6th ed. New York: John Wiley & Sons, Inc., 2010.

Levy, Stuart B. "The Challenge of Antibiotic Resistance." *Scientific American* 278 (March 1998): 46–53.

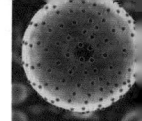

Lou, H. and R. F. Gagel. "Alternative RNA Processing—Its Role in Regulating Expression of Calcitonin/Calcitonin Gene-related Peptide." *Journal of Endocrinology* 156 (1998): 401–405.

Macklis, Roger M. "The Great Radium Scandal." *Scientific American* 269 (August 1993): 94–99.

Mair, William. "How Normal Cells Can Win the Battle for Survival Against Cancer Cells." *PLoS Biology* 8 (July 2010): e1000423. Available online. URL: http://www.plosbiology.org. Accessed July 14, 2010.

Maugh II, Thomas H. and Shari Roan. "Artificially Created Cell Called a Scientific Feat." *Los Angeles Times*, May 20, 2010.

Medawar, P.B. and J.S. Medawar. *Aristotle to Zoos*. Cambridge, Mass.: Harvard University Press, 1983.

Mildenhall, D.C. "*Hypericum* Pollen Determines the Presence of Burglars at the Scene of a Crime: An Example of Forensic Palynology." *Forensic Science International* 163 (November 22, 2006): 231–235.

Mildenhall, D.C., P.E.J. Wiltshire, and V.M. Bryant. "Forensic Palynology: Why Do It and How It Works." *Forensic Science International* 163 (November 22, 2006): 163–172.

Mosley, Amber L., John A. Corbett, and Sabire Özcan. "Glucose Regulation of Insulin Gene Expression Requires the Recruitment of p300 by the β-Cell-Specific Transcription Factor Pdx-1." *Molecular Endocrinology* 18 (2004): 2279–2290.

Murray, Robert K., Daryl K. Granner, and Victor W. Rodwell. *Harper's Illustrated Biochemistry,* 27th ed. New York: Lange Medical Books/McGraw-Hill, 2006.

Naik, Gautam. "Scientists Build a Rat Lung." *Wall Street Journal*, June 25, 2010.

Nicholls, Henry. "The Chemistry Set Generation." *Chemistry World* (December 2007): 42–45.

Poliseno L., L. Salmena, J. Zhang, B. Carver, W.J. Haveman, and P.P. Pandolfi. "A Coding-independent Function of Gene and Pseudogene mRNAs Regulates Tumour Biology." *Nature*. 465 (2010): 1033–1038.

The Revigator Water Jar for Every Home. San Francisco: The Revigator Water Jar Company, 1928.

Rosselló, Ricardo A. and David H. Kohn. "Cell Communication and Tissue Engineering." *Communicative & Integrative Biology* 3 (2010): 53–56.

Rosselló, Ricardo A., Zhuo Wang, Eddy Kizana, Paul H. Krebsbach, and David H. Kohn. "Connexin 43 as a Signaling Platform for Increasing the Volume and Spatial Distribution of Regenerated Tissue." *Proceedings of the National Academy of Sciences USA* 106 (2009): 13219–13224.

Ruddon, Raymond W. *Cancer Biology,* 4th ed. New York: Oxford University Press, Inc., 2007.

Saey, Tina Hesman. "To Catch a Thief, Follow His Filthy Hands." *Science News* 177 (April 10, 2010): 13.

———. "Neandertal Genome Yields Evidence of Interbreeding with Humans." *Science News* (June 5, 2010): 5.

———. "Cancer's Little Helpers." *Science News* 178 (August 28, 2010): 18–21.

Sanders, Laura. "Genome From a Bottle Turns One Bacterium Into Another." *Science News* 177 (June 19, 2010): 5–6.

Severs, Nicholas J., Steven R. Coppen, Emmanuel Dupont, Hung-I Yeh, Yu-Shien Ko, Tsutomu Matsushita. "Gap Junction Alterations in Human Cardiac Disease." *Cardiovascular Research* 62 (2004): 368–377.

Smith, Roger S. and Barbara H. Iglewski. "*Pseudomonas aeruginosa* Quorum Sensing as a Potential Antimicrobial Target." *Journal of Clinical Investigation* 112 (2003): 1460–1465.

Sompayrac, Lauren. *How the Immune System Works,* 3rd ed. Malden, Mass.: Blackwell Publishing, 2008.

Tuppen, Helen A.L., Emma L. Blakely, Douglass M. Turnbull, and Robert W. Taylor. "Mitochondrial DNA Mutations and Human Disease." *Biochim Biophys Acta* 1797 (2010): 113–128.

Van Delden, Christian and Barbara H. Iglewski. "Cell-to-Cell Signaling and *Pseudomonas aeruginosa* Infections." *Emerging Infectious Diseases* 4 (1998): 551-560.

vonHoldt, Bridgett M., John P. Pollinger, Kirk E. Lohmueller, Eunjung Han, Heidi G. Parker et al. "Genome-wide SNP and Haplotype Analyses Reveal a Rich History Underlying Dog Domestication." *Nature* 464 (2010): 898–902.

Williams, Paul. "Quorum Sensing: An Emerging Target for Antibacterial Chemotherapy?" *Expert Opinion on Therapeutic Targets* 6 (2002): 257–274.

Wolpert, Lewis. *How We Live & Why We Die: The Secret Lives of Cells.* New York: W.W. Norton & Company, 2009.

Wong, Kate. "Our Inner Neandertal." *Scientific American* 303 (July 2010): 18–20.

Yeager, Mark. "Structure of Cardiac Gap Junction Intercellular Channels." *Journal of Structural Biology* 121 (1998): 231–245.

Further Resources

Books

Boudreau, Gloria. *The Immune System*. San Diego: KidHaven Press, 2004.

Bozzone, Donna M. *Cancer Genetics*. New York: Chelsea House, 2007.

Bozzone, Donna M. *Causes of Cancer*. New York: Chelsea House, 2007.

Claybourne, Anna. *Introduction to Genes and DNA*. London: Usborne Publishing Limited, 2003.

Fridell, Ron. *Decoding Life: Unraveling the Mysteries of the Genome*. Minneapolis: Lerner Publishing Group, 2004.

Laberge, Monique. *Biochemistry*. New York: Chelsea House, 2008.

Phelan, Glen. *Double Helix*. Washington, D.C.: National Geographic Children's Books, 2006.

Simpson, Kathleen. *National Geographic Investigates: Genetics: From DNA to Designer Dogs*. Washington, D.C.: National Geographic Children's Books, 2008.

Stewart, Gregory J. *The Immune System*. New York: Chelsea House, 2009.

Web sites

CELLS Alive!
http://www.cellsalive.com

This Web site offers animated cell models, and animations of mitosis and meiosis. Cell enthusiasts can also view time-lapse images of cultured human and bacterial cells via webcam.

Genetic Science Learning Center
http://learn.genetics.utah.edu

The University of Utah provides a wide range of information about various cell activities. Readers can peruse colorful presentations on basic functions of DNA, RNA, proteins, protein synthesis, cellular membranes, and organelles, as well as cell communication.

The Official Web Site of the Noble Prize
http://nobelprize.org/educational/medicine/immunity/immune-detail.html

The Nobel Prize Web site offers a tutorial, "The Immune System–in More Detail," which provides an overview of the immune system.

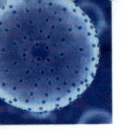

Online Biology Book

http://www.emc.maricopa.edu/faculty/farabee/biobk/biobooktoc.html

This electronic book has chapters that cover cell organization, metabolism, communication, immunology, mitosis, meiosis, and many other topics about cells.

Science Daily

http://www.sciencedaily.com

The Science Daily Web site is a comprehensive source for news about science research, including research into cells, DNA, cell communication, immunology, and cancer.

Picture Credits

Index

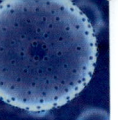

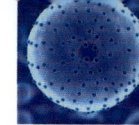

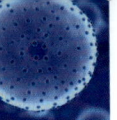

About the Author

Phill Jones earned a Ph.D. in Physiology/Pharmacology from the University of California, San Diego. After completing postdoctoral training at Stanford University School of Medicine, he joined the Department of Biochemistry at the University of Kentucky Medical Center as an assistant professor. Here, he taught topics in molecular biology and medicine, and researched aspects of gene expression. He later earned a JD at the University of Kentucky College of Law and worked ten years as a patent attorney, specializing in biological, chemical, and medical inventions. Dr. Jones is now a full-time writer. His articles have appeared in *Today's Science on File*, *The World Almanac and Book of Facts*, *History Magazine*, *Forensic Magazine*, *Genomics and Proteomics Magazine*, *Encyclopedia of Forensic Science*, *The Science of Michael Crichton*, *Forensic Nurse Magazine*, *Nature Biotechnology*, *Information Systems for Biotechnology News Report*, *Law and Order Magazine*, *PharmaTechnology Magazine*, and educational testing publications. His books *Sickle Cell Disease* (2008) and *The Genetic Code* (2010) were published by Chelsea House. He also writes and teaches an online course in forensic science for writers.